Math Mammoth
The Four Operations
(with a Touch of Algebra)

By Maria Miller

Contents

Introduction

Math Mammoth The Four Operations (with a Touch of Algebra) is a mathematics worktext meant primarily for fifth and sixth grades. Being a worktext means that this book is a textbook and workbook together: the lessons include both the explanations of the concepts, as well as practice exercises.

The lessons in this worktext have been taken from the Math Mammoth complete curriculum for fifth and sixth grades. For this reason, they may not always flow smoothly from one lesson to the next with a perfect continuity, though I have tried to present them here in the most logical order.

The main topics of this book are mental math strategies, simple equations and expressions, the order of operations, partial products, using all four operations in solving problems, and number patterns in the coordinate grid.

Students encounter the exact definition of an *equation* and an *expression*. They practice the order of operations with problems that also reinforce the idea of the equal sign ("=") as denoting equality of the right and left sides of an equation. These kinds of exercises are needed because students may think that an equal sign signifies *the act of finding the answer* to a problem (such as $134 + 23 = ?$), which is not so.

Students solve simple equations with the help of a bar model. They also use the pan balance model to solve simple one and two-step equations and even some three-step equations.

We review the thought of partial products (multiplying and dividing in parts), including how those partial products are seen in the multiplication algorithm, and how to use mental math to divide numbers in parts.

Although the book is named "The Four Operations," the idea is not to practice each of the four operations separately, but rather to see how they are used together in solving problems and in simple equations. We are trying to develop the student's *algebraic thinking*, including the abilities to: translate problems into mathematical operations, comprehend the many operations needed to yield an answer to a problem, "undo" operations, and so on. Many of the ideas in this book are preparing students for algebra.

We also offer free resources on the Math Mammoth website that can complement the lessons in the book:

- Check out the free videos at https://www.mathmammoth.com/videos
  The videos for 5th grade, chapter 1 (also chapter 3 & 5), and 6th grade, chapter 1, best match the lessons in this book.

- At https://www.mathmammoth.com/practice you will find free online activities and games to practice various math concepts.

I wish you success in teaching math!

Maria Miller, the author

Helpful Resources on the Internet

We have compiled a list of external Internet resources that match the topics in this book. This list of links includes web pages that offer:

- **online practice** for concepts;

- online **games**, or occasionally, printable games;

- **animations** and interactive **illustrations** of math concepts;

- **articles** that teach a math concept.

We heartily recommend you take a look at the list. Many of our customers love using these resources to supplement the bookwork. You can use the resources as you see fit for extra practice, to illustrate a concept better, and even just for some fun. Enjoy!

https://l.mathmammoth.com/blue/fouroperations

SCAN ME

Warm-up: Mental Math

Add in parts.	Use rounded numbers, then correct the error.
57 + 34 = ? Add the tens: 50 + 30 = 80. Add the ones: 7 + 4 = 11. Lastly, add the two sums: 80 + 11 = 91.	29 + 18 = ? 29 is close to 30, and 18 is close to 20. 30 + 20 = 50. But that is 3 too many, so the correct answer is 47.
Subtract in parts.	**Use rounded numbers, then correct the error.**
81 − 34 = ? Subtract 30 first: 81 − 30 = 51. Then subtract four: 51 − 4 = 47.	75 − 39 = ? 39 is close to 40, so subtract 75 − 40 = 35. You subtracted one too many, so add one to get the correct answer 36.

1. Add and subtract using the tricks explained above.

a.	b.	c.
19 + 19 = _____ 28 + 47 = _____	19 + 19 + 57 = _____ 44 + 12 + 29 = _____	100 + 200 + 2,000 + 5,500 = _____ 400 + 12,000 + 5,000 + 320 = _____
d.	**e.**	**f.**
33 − 17 = _____ 81 − 47 = _____	34 − 19 + 12 = _____ 85 − 12 + 55 = _____	1,500 − 250 − 250 = _____ 400 − 7 − 40 − 100 = _____

2. A track has four legs of different lengths: (a) 1 km 200 m, (b) 700 m, (c) 1 km 500 m, and (d) 900 m. What is the total length of the track?

Hint: "kilo" in kilometer refers to one thousand.

3. A cold front just arrived, and the temperature dropped 14 degrees. It is now 74°F. How hot was it before?

4. Four crates of apples weigh a total of 56 kg. The first one weighs 12 kg, the second one 15 kg, and the third one 22 kg. Find the weight of the fourth crate of apples.

5. Solve in your head.

a. 127 + _____ = 200	**b.** 250 + _____ + 300 = 760	**c.** _____ − 34 = 56

7

6. Multiply.

a. $20 \times 6 =$ _____	**b.** $10 \times 35 =$ _____	**c.** $400 \times 500 =$ _____
$200 \times 6 =$ _____	$100 \times 35 =$ _____	$60 \times 80 =$ _____
$200 \times 600 =$ _____	$20 \times 100 =$ _____	$100 \times 430 =$ _____

7. Continue the patterns for the next five numbers.

   a. 60, 120, 180, 240, ...

   b. 1,080, 960, 840, 720, ...

   c. 130, 170, 210, 250, ...

8. Estimate the cost of buying two skirts for $26.95
   and three pairs of socks for $3.29 each.
   (Use rounded numbers.)

Multiply part-by-part	**5 times a number**
Multiply ones, tens, and hundreds separately. Add.	Find 10 times half of the number.
$3 \times 62 = \underline{3 \times 60} + \underline{3 \times 2} = 186$	$5 \times 28 = \underline{10 \times 14} = 140.$
9 times a number	**11 times a number**
Find 10 times a number and subtract that number once.	Find 10 times the number, and then add that number.
$9 \times 55 = \underline{10 \times 55 - 55}$ $= 550 - 55 = 495$	$11 \times 38 = \underline{10 \times 38 + 38}$ $= 380 + 38 = 418$

9. Multiply using the "tricks" explained above.

   a. $5 \times 26 =$ _____   **b.** $5 \times 43 =$ _____   **c.** $6 \times 41 =$ _____

   d. $5 \times 107 =$ _____   **e.** $9 \times 15 =$ _____   **f.** $9 \times 32 =$ _____

   g. $7 \times 205 =$ _____   **h.** $3 \times 211 =$ _____   **i.** $11 \times 25 =$ _____

   j. $11 \times 18 =$ _____   **k.** $4 \times 32 =$ _____   **l.** $9 \times 109 =$ _____

The Order of Operations

Mathematicians have decided that if there are many operations, they are to be done in a certain order. This is to prevent confusion.

1. First solve whatever is inside parentheses.

Parentheses mark what operations are priorities to be done first.

2. Next, solve multiplications and divisions, from left to right.

This does not mean multiplications are to be done before divisions. Instead, they are all equally important, or "on the same level". For example, in $45 \div 5 + 2 \times 8$, do both the division and the multiplication first, before the addition. (It won't matter whether you divide or multiply first.)

If there are several multiplications and divisions in a row (without addition or subtraction in between), do them from left to right. For example, in $36 \div 9 \times 5$, solve $36 \div 9$ first.

3. Last, solve additions and subtractions, from left to right.

Again, this doesn't mean additions are done before subtractions. Instead, they're to be done from left to right. For example, in $200 - 50 + 30 + 7$, solve $200 - 50$ first.

1. Solve what is in the parentheses first. You can enclose the operation to be done first in a "bubble."

Example 1.	a. $(50 - 2) \div (3 + 5)$	b. $20 \times (1 + 7 + 5)$
$(36 + 4) \div (5 + 5)$ $= 40 \div 10$ $= 4$	c. $2 \times (600 \div 60) + (19 - 8)$	d. $180 \div (13 - 7 + 3)$

2. Solve. When there are several multiplications and divisions in a row, do them from left to right.

Example 2.	a. $36 \div 4 \div 3$	b. $1{,}200 \div 4 \times 5 \div 3$
$24 \div 3 \times 2 \div 4$ $= 8 \times 2 \div 4$ $= 16 \div 4 = 4$	c. $7 \times 90 \div 2 \times 2 \div 10$	d. $5 \times 6 \div 3 \div 2 \times 20$

Parentheses are used to change the normal order of operations. For example, if we want 9 and 18 added first, then the result multiplied by 3, we write $3 \times (9 + 18)$ or $(9 + 18) \times 3$.

(What would get done first if you wrote $3 \times 9 + 18$ or $9 + 18 \times 3$?)

3. Write a calculation for the following, and solve.

   a. First subtract 9 from 30, then multiply the result by 5.

   b. First multiply 7 and 6, then add 20 to the result.

   c. First add 14, 15, and 16, then divide that by 3.

   d. First add 27 and 37, then subtract what you get from 100.

4. Now let's do it with more operations.

   a. First add 26 and 6, then multiply that by 2, and lastly subtract what you got thus far, from 90.

   b. First multiply 5 and 7, subtract the result of that from 100, and lastly add 34 to it.

   c. First divide 36 by 9, multiply the result by 5, and subtract that from 55.

5. Solve in the right order. You can enclose the operation to be done first in a "bubble" or a "cloud."

a. $(8 + 16) \div 3 \div 2 =$ _____	**b.** $10 + 2 \times 9 + 8 =$ _____
c. $25 + 8 \times 5 \div 2 =$ _____	**d.** $10 + 2 \times (9 + 8) =$ _____
e. $120 - 2 \times (11 - 5) =$ _____	**f.** $2 \times (100 - 80 + 20) =$ _____

6. Division can also be written with a fraction line. Solve in the right order.

a. $6 + \dfrac{24}{2} =$ _____	**b.** $40 + \dfrac{32}{2} - 6 =$ _____	**c.** $\dfrac{54}{6} - 3 \times 2 =$ _____

Equations

> An **expression** contains numbers, letters, and operation symbols—but no equals sign.
> For example, "$40 \times 2 + 6 \times 5$" is an expression. A single number or letter, such as 9,
> is also an expression.
>
> An **equation** contains two expressions separated by an equals sign, "=".
> Here are two examples: $24 = 11 + 13$ and $2x - 13 = 6/y$. Even $0 = 0$ is an equation!

1. Equation or expression? (Do not solve these.)

    a. $4t = 180$ **b.** $2 + 60 \times 345 \div 9$ **c.** $15 = x + y$

    d. $\dfrac{5.4 - 2.12}{0.4} = 8.2$ **e.** $1{,}000 = 1{,}000$ **f.** $12 - \dfrac{24 \div 0.8}{189}$

2. Which expression matches each problem? Also, solve the problems.

a. Mark bought three light bulbs for $8 each and paid with $50. What was his change?	**(1)** $3 \times \$8 - \$50$ **(2)** $\$50 - \$8 + \$8 + \$8$	**(3)** $\$50 - 3 \times \$8$ **(4)** $\$50 - (\$8 - \$8 - \$8)$
b. Shirts that cost $16 each are discounted by $5, so Mom bought six of them. What was the total cost?	**(1)** $\$16 - \$5 \times 6$ **(2)** $6 \times (\$16 - \$5)$	**(3)** $\$16 \times 6 - \$5$ **(4)** $(\$16 - 6) \times 5$
c. Andy bought a salad for $8 and a pizza for $13, and shared the cost evenly with his friend. What was Andy's share of the cost?	**(1)** $\$8 + \$13 \div 2$ **(2)** $\$2 \div (\$8 + \$13)$ **(3)** $2 \times \$8 + 2 \times \$13$ **(4)** $(\$8 + \$13) \div 2$	
d. Melissa shared equally the cost of a meal with three other people, and the cost of a taxi with two other people. The meal cost $48 and the taxi cost $30. How much did Melissa pay?	**(1)** $\$48 \div 4 + \$30 \div 3$ **(2)** $(\$48 + \$30) \div 3 \div 2$ **(3)** $\$48 \div 3 + \$30 \div 2$ **(4)** $(\$48 + \$30) \div 5$	

An **equation** contains <u>two expressions</u>, separated by an equals sign.

$$120 - 75 \;\; = \;\; 3 \times 15$$

This is the left side of the equation. This is the right side of the equation.

If the left and right sides have the same value, it is a **true equation**. If not, it is a **false equation**.

The equation below is false.

$$4 + 5 \;\; = \;\; 21 - 3$$

left side   right side

$$18 \;\; = \;\; x - 3$$

Solving an equation means finding the value of the **unknown** (x) that makes it true.

The value $x = 21$ makes this equation true, so we say $x = 21$ is the solution.

3. If the equation is false, change one number in it to make it true.

a. $6 + \dfrac{32}{8} = 5$	**b.** $(6 - 2) \times 3 = 5 + 5$	**c.** $5 \times 2 = 16 \div 2 + 2$

4. Place parentheses into these equations to make them true.

a. $10 + 40 + 40 \times 2 = 180$	**b.** $144 = 3 \times 2 + 4 \times 8$	**c.** $40 \times 3 = 80 - 50 \times 4$

5. Find a number to fit in the box so the equation is true.

a. $40 = (\boxed{} + 9) \times 2$	**b.** $4 \times 8 = 5 \times 6 + \boxed{}$	**c.** $4 + 5 = (20 - \boxed{}) \div 2$
d. $81 = 9 \times (2 + \boxed{})$	**e.** $\boxed{} \times 11 = 12 + 20 \times 6$	**f.** $(4 + 5) \times 3 = \boxed{} \div 2$

6. Solve these simple equations.

a. $s \times 2 = 660$ $s = $ _____	**b.** $\dfrac{x}{2} = 5$ $x = $ _____	**c.** $200 - y = 60$ $y = $ _____

7. Build at least three true equations using only the symbols and numbers given. You may use the same number or symbol many times.

$$11, 3, 1, -, +, \times, (\ ), =$$

Review: Addition and Subtraction

Addition has to do with many parts and their total. You add the parts to get the total.

In this addition, one of the parts is unknown.

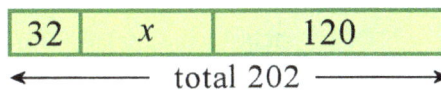

$$32 + x + 120 = 202$$

Subtraction also has to do with a total and parts. A subtraction equation <u>starts with the total</u>.

We can write several subtractions to match this bar model. In each subtraction, we subtract two of the three parts from the total, and the answer is the remaining part.

Which subtraction on the right can be used to find (or solve) the unknown x?

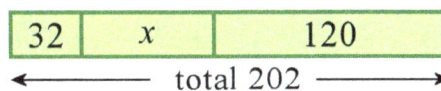

$$202 - 32 - x = 120$$

$$202 - 120 - 32 = x$$

$$202 - x - 120 = 32$$

1. Write one addition equation *and* one subtraction equation to match each bar model. Then solve for x.

a. Addition:

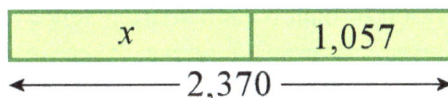

Subtraction:

$x =$ _____

b. Addition:

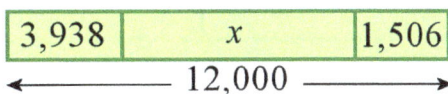

Subtraction:

$x =$ _____

c. Addition:

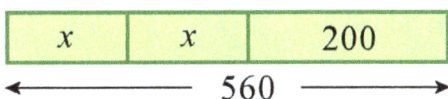

Subtraction:

$x =$ _____

Sum and addends	**Minuend, subtrahend, and difference**

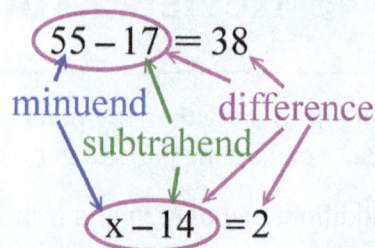

$$5 + 8 + 13 = 26$$

addends   sum

$$x + 20 = 70$$

$$55 - 17 = 38$$

minuend   difference

subtrahend

$$x - 14 = 2$$

The numbers that are being added are called **addends**. The result is a **sum**—even if you haven't yet calculated it. So "5 + 8" is called a *sum*.

Examples:

"8 + 13" is a sum. 8 and 13 are the addends.

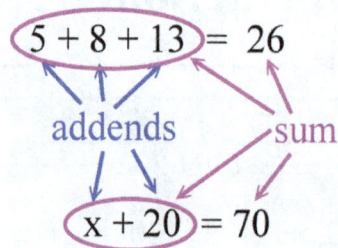

"20 + x" is a sum: It is the sum of 20 and x.

You can call "5 + 8 + 13" the *sum written*, and the answer 26 you can call the sum that has been *solved* or *calculated*.

The number that you subtract from is called the **minuend**. The number you subtract is the **subtrahend**. (The minuend comes before the subtrahend, just like "M" comes before "S" in the alphabet.)

The result is the **difference**, even if it hasn't yet been solved. So "55 − 17" is a *difference*.

Examples:

The difference of 55 and 17 is written as 55 − 17. We can solve or calculate that to get 38.

The difference of x and *14* is written as x − 14.

2. Write an expression *or* an equation to match each written sentence.

a. The sum of 68 and s	**b.** The difference of y and 37
c. The sum of 60, b, and 40 equals 120.	**d.** The difference of 80 and x is 35.

3. Match the written expressions with the number expressions. (You don't have to solve these.)

    a. The sum of 7 and 5 is subtracted from 20. $(7 - 5) + 20$

    b. The difference of 7 and 5 is subtracted from 20. $20 - 5 - 7$

    c. 20 is added to the difference of 7 and 5. $7 + (20 - 5)$

    d. The difference of 20 and 5 is added to 7. $20 - (7 + 5)$

                                                                             $20 - (7 - 5)$

4. Write an expression. Consider the order of operations!

a. The difference of 15 and 6 is added to 16.	**b.** The sum of 5 and 80 is subtracted from 100.

> **Remember?** Whether you subtract a sum of several numbers → $100 - (40 + 20 + 30) = 10$
>
> or subtract the numbers one by one → $100 - 40 - 20 - 30 = 10$
>
> ...the answer is the same!

5. Solve. Notice: some problems have the same answer. Which ones?

a. $7,000 - (1,500 + 2,500) =$ _____	**b.** $600 + 30 - 30 + 30 - 30 =$ _____
$7,000 - 2,500 - 1,500 =$ _____	$600 - (30 + 30 + 30 + 30) =$ _____
$7,000 - (2,500 - 1,500) =$ _____	$600 - 30 - 30 - 30 - 30 =$ _____

Which number sentence matches the problem? *You don't have to calculate the answer.*

6. Mark bought 14 wheelbarrows for $58 each, and paid with $900. What was his change?

    a. $14 \times \$58 - \$900$ **b.** $\$900 - 14 \times \$58$

    c. $\$900 \times 14 \times \$58$ **d.** $\$58 - \$900 \times 14$

7. Sarah baked three cakes. On top of each, she put 24 chocolate chips and 12 banana slices. How many total items are on top of the cakes?

    a. $24 + 12 \times 3$ **b.** $24 + 24 + 12 + 12 + 12$

    c. $3 \times (24 + 12)$ **d.** $3 \times 24 + 12$

8. Jack and Jill bought 9 toys for $7 each, and shared the cost equally. How much did each pay?

    a. $\$7 + 9 \times 2$ **b.** $9 \times \$7 \div 2$

    c. $9 \times 2 \times \$7$ **d.** $\$7 \div 9 \times 2$

9. Write a single expression using numbers and operations for each problem, not just the answer!

a. You bought 15 toy cars for $2 each and a sand toy set for $6. You paid with $50. What was the total cost?   What was your change?
b. Three children bought strawberries for $9, ice cream for $8, and cheese for $13. They shared the cost equally. How much did each child pay?
c. The price of a phone that costs $128 is lowered (discounted) by $31. George bought five of them. What was the total cost?

Review: Multiplication and Division

Multiplication and division of whole numbers have to do with things or groups of the same size.

When you multiply the number of groups by the amount in each group (or the other way around), you get the <u>total</u>.

When you divide the total by the number of groups, you get the amount in each group.

When you divide the total by the amount in each group, you get the number of groups.

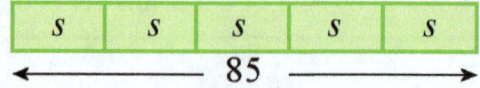

s	s	s	s	s

$\longleftarrow$ 85 $\longrightarrow$

We can write four equations to match the model:

$$5 \times s = 85 \qquad 85 \div 5 = s$$

$$s \times 5 = 85 \qquad 85 \div s = 5$$

1. Write one multiplication equation and one division equation for each bar model. Then solve for w.

a.
305	305	305	305

$\longleftarrow$ w $\longrightarrow$

_____ × _____ = _____

_____ ÷ _____ = _____

$w =$ _____

b.
w	w	w	w	w

$\longleftarrow$ 305 $\longrightarrow$

_____ × _____ = _____

_____ ÷ _____ = _____

$w =$ _____

2. Which equation matches which bar model? Also, solve for y.

<u>Equations:</u>

$6 \times y = 90$

$y \div 6 = 90$

a.
y	y	y	y	y	y

$\longleftarrow$ 90 $\longrightarrow$

$y =$ _____

b.
90	90	90	90	90	90

$\longleftarrow$ y $\longrightarrow$

$y =$ _____

3. Draw a bar model to represent each equation, and solve the equation.

a. $R \div 5 = 120$	b. $5 \times R = 120$

c. $y \div 12 = 60$

Product and factors	**Dividend, divisor, and quotient**
$5 \times 6 \times 3 = 90$ factors → product $s \times 12 = 96$	dividend quotient $\dfrac{x}{20} = 5$ divisor
The numbers that are being multiplied are called **factors**. The result is called a **product**—even if you have not yet calculated it. So "5 × 6" is called a product.	The number you divide is called the **dividend**. The number you divide by is the **divisor**. The result is the **quotient**, even if it has not yet been solved. So "$x \div 20$" is a quotient (of x and 20).
Examples: 5×6 is a product. 5 and 6 are the factors. $s \times 12$ is a product: it is the product of s and 12. You can call $5 \times 6 \times 3$ the *product written,* and the answer 90 you can call the product that has been *solved* or *calculated.*	**Examples:** The quotient of 100 and 5 is written as $100 \div 5$, or using the fraction line as $\dfrac{100}{5}$. We can solve or calculate that to get 20. The quotient of x and 20 is written $x \div 20$ or $\dfrac{x}{20}$.

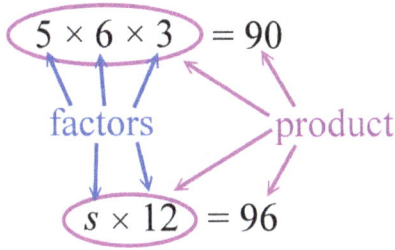

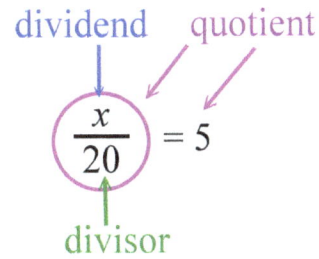

4. Write an expression or an equation to match each written sentence.

a. The product of 52 and 8	**b.** The quotient of 15,000 and 300
c. The product of 4, S, and 18	**d.** The quotient of 80 and x
e. The quotient of 240 and 8 is 30	**f.** The product of 3, 5, and T is 60

5. Write a division equation where the dividend is 280, the quotient is 4, and the divisor is unknown. Use a letter for the unknown. Then find the value of the unknown.

6. Write a division equation where the quotient is 3, the divisor is 91, and the dividend is unknown. Use a letter for the unknown. Then find the value of the unknown.

Look carefully at this expression: $3 \times 47 + 8 \times 47$. Think of it as three copies of 47, and another eight copies of 47. In total, we have 11 copies of 47, or 11×47.

Similarly, $9 \times 165 - 4 \times 165$ is like saying that we have 9 copies of the number 165, and we take away four copies of that number. What is left? Five copies of that number, or 5×165.

7. For each two expressions, decide if the answers are the same or not. Do *not* calculate the answers.

a. $3 \times 417 - 417$ 2×417	**b.** $6 \times 799 - 2 \times 799$ 3×799	**c.** $389 + 389 + 389 + 72 + 72 + 72$ $3 \times 389 + 3 \times 72$
d. 16×68 $9 \times 68 + 7 \times 68$	**e.** $500 - 25 + 19$ $500 - (25 + 19)$	**f.** $832 - 225 - 195$ $832 - (225 + 195)$

8. Which number sentence matches the problem? *You don't have to calculate the answer.*

The sides of a rectangular park measure 26 ft and 43 ft. Ashley ran around it three times. What is the distance she ran?

a. $(26 + 43) \times 3$ **b.** $3 \times 2 \times (26 + 43)$

c. $26 + 43 + 26 + 43$ **d.** $3 \times 26 + 43 + 26 + 43$

9. Look at the division equations. In each, the *dividend* is the unknown. Explain how you can find the unknown. (You don't have to actually solve the equations; just explain *how* to solve them.)

$x \div 5 = 4$ $N \div 12 = 60$

$y \div 8 = 100$ $M \div 83 = 149$

10. Look at the division equations. In each, the *divisor* is the unknown. Explain how you can find the unknown. (You don't have to actually solve the equations; just explain *how* to solve them.)

$16 \div x = 8$ $350 \div N = 50$

$72 \div y = 9$ $120 \div M = 6$

11. Solve for the unknown N or M.

a. $5 \times M = 20$	**b.** $M \div 3 = 5$	**c.** $45 \div M = 5$
d. $4 \times N = 8,800$	**e.** $N \div 20 = 600$	**f.** $640 \div N = 80$

Balance Problems and Equations 1

Here you see a pan balance, or scales, and some things on both pans. Each rectangle represents an unknown (and "weighs" the same, or has the same value).

Since the balance is *balanced* (neither pan is going down—they are level with each other), the two sides (pans) of the scales weigh the <u>same</u>.

This portrays a mathematical equation: what is in the left pan <u>equals</u> what is in the right pan. (Things in the same pan are simply added.)

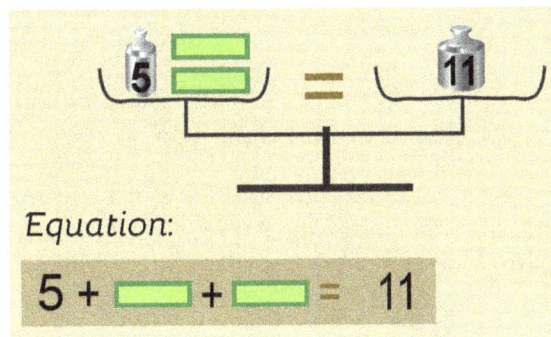

Equation:

$$5 + \boxed{} + \boxed{} = 11$$

The equation is:

$$5 + \boxed{} + \boxed{} = 11$$

(If it helps you, you can think of kilograms or pounds.)

When we figure out how much the unknown shape weighs, we solve the equation.

The solution is: $\boxed{} = 3$

1. Write an equation for each balance. Then use mental math to solve how much each geometric shape "weighs." You can write a number inside each of the geometric shapes to help you.

a.

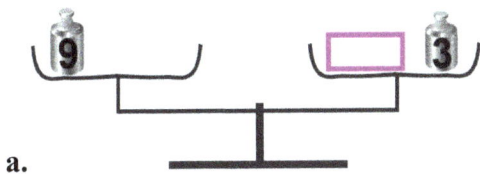

Equation: $9 = \boxed{} + 3$

Solution: $\boxed{} = 6$

b.

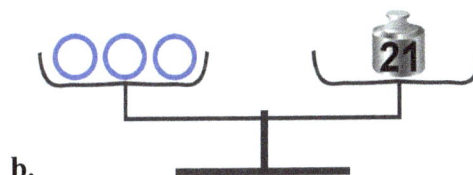

Equation:

_____

Solution: $\bigcirc = $ _____

c.

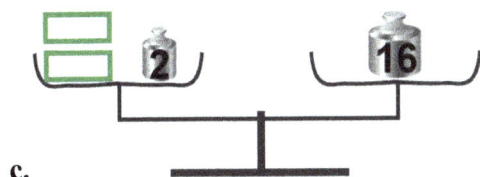

Equation:

_____

Solution: $\boxed{} = $ _____

d.

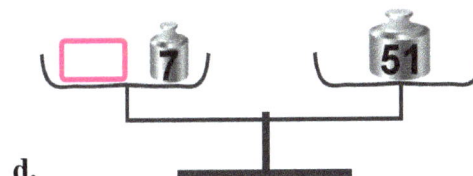

Equation:

_____

Solution: $\boxed{} = $ _____

19

From now on we will use *x* for the unknown instead of a geometric shape. It is the most commonly used letter of the alphabet to signify an unknown.

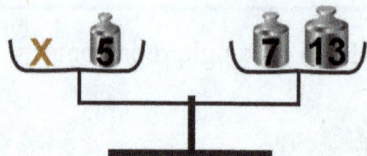

$$x + 5 = 7 + 13$$
$$x + 5 = 20$$
$$x = 15$$

Example 1. To solve this equation, first add 7 and 13 that are in the right "pan".

We get $x + 5 = 20$. The solution is easy to see now with mental math: $x = 15$. You can also use subtraction: $x = 20 - 5 = 15$.

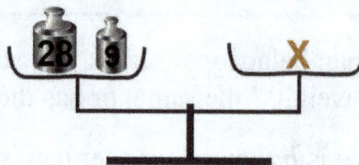

$$28 + 9 = x$$
$$37 = x$$
$$x = 37$$

Example 2. Sometimes *x* is on the right side of the equation. That is not a problem. In the last step you can flip the sides, so that your last line will be $x = $ (something).

Notice that we *align the equal signs* when solving an equation. It keeps everything tidy and easy to read.

2. Write an equation. Write a second step if necessary. Lastly write what *x* stands for.

a.

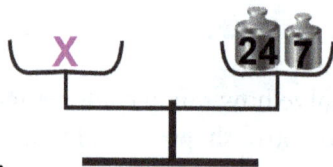

_____ = _____

x = _____

b.

_____ = _____

_____ = _____

x = _____

3. Draw *x*'s and weights on the left and right sides on the two pans to match the given equation, then solve. You may not need all the empty lines provided.

a.

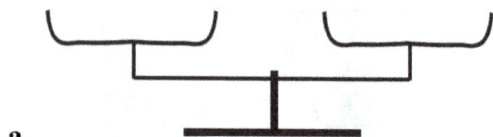

$$x + 18 = 5 + 31$$

_____ = _____

_____ = _____

_____ = _____

b.

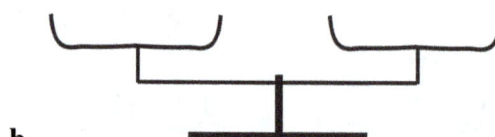

$$8 + 17 = 11 + x$$

_____ = _____

_____ = _____

_____ = _____

20

Whenever there are lots of x's in the same pan, use this shorthand notation:

- $x + x$ is written as $2x$. It means 2 times x.
- $x + x + x$ is written as $3x$. It means 3 times x.
- $x + x + x + x$ is written as $4x$, and so on.

We simply omit the multiplication sign between a number and a letter (such as 4 and x).

Example 3.

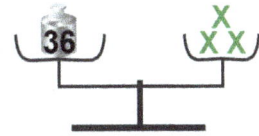

You can use *division* to solve this.

$$36 = 3x$$
$$12 = x$$

Lastly, flip the sides. → $x = 12$

4. Write an equation to match the balance. Then solve what x stands for.

a.

_____ = _____

_____ = _____

x = _____

b.

_____ = _____

_____ = _____

x = _____

c.

_____ = _____

_____ = _____

x = _____

5. Draw x's and weights on the left and right sides on the two pans to match the given equation, then solve. You may not need all the empty lines provided.

a.

$$3x = 16 + 35$$

_____ = _____

_____ = _____

_____ = _____

b.

$$2 + 27 + 25 = 6x$$

_____ = _____

_____ = _____

_____ = _____

Puzzle Corner

Solve the equations.

a. $3{,}928 + 3{,}943 = 17x$

b. $10{,}000 - 5{,}493 - 834 - 3{,}673 = 22x$

Balance Problems and Equations 2

<table>
<tr>
<td>If there are x's on both sides, use this "trick": Take away, or subtract, the same amount of x's from both sides so that you will only have ONE x left on one side.</td>
<td>

Example 1.

$2x = x + 34$

$x = 34$

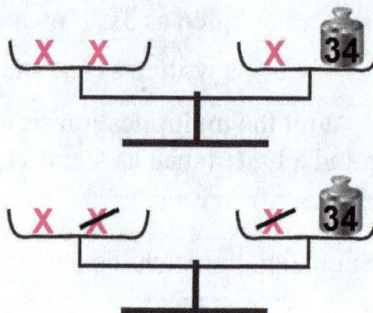
</td>
</tr>
</table>

1. First write the equation as the balance shows it. Then solve, crossing out x's from both sides.

a.

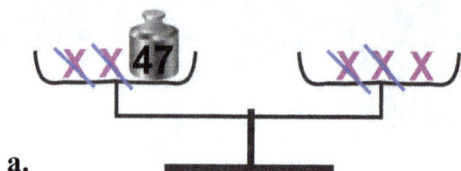

$2x + 47 = 3x$

$47 = x$

$x = 47$

b.

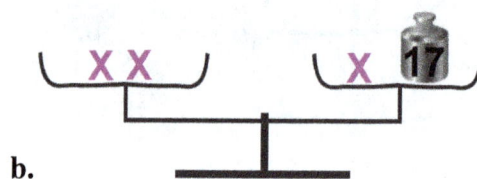

_____ = _____

_____ = _____

c.

_____ = _____

_____ = _____

_____ = _____

d.

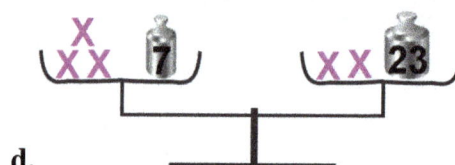

_____ = _____

_____ = _____

_____ = _____

e.

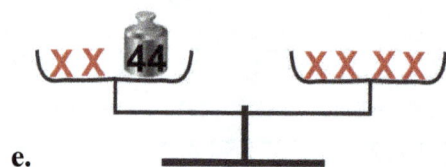

_____ = _____

_____ = _____

_____ = _____

_____ = _____

f.

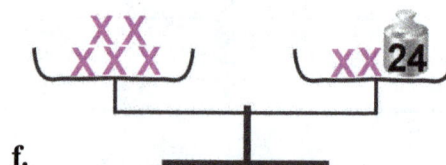

_____ = _____

_____ = _____

_____ = _____

You can also remove the same amount of "weight" from both sides.

Here, it helps to remove, or subtract, 6 kg from both sides.

You can indicate that by crossing out the 6-kg weight, and by crossing out the number 36 on the other weight and writing 30 in its place.

Example 2.

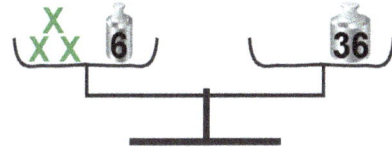

$$3x + 6 = 36$$
$$3x = 30$$
$$x = 10$$

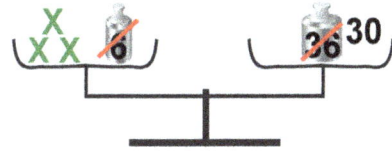

Example 3.

Can you follow the solution on the right, and cross out items from the pans accordingly?

$$3x + 9 = x + 27$$
$$3x = x + 18$$
$$2x = 18$$
$$x = 9$$

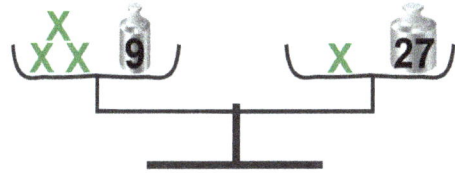

2. Solve.

a.

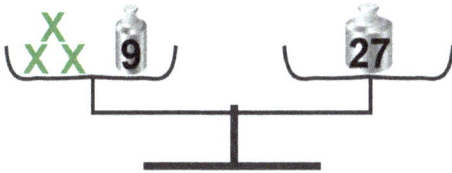

_____ = _____

_____ = _____

_____ = _____

b.

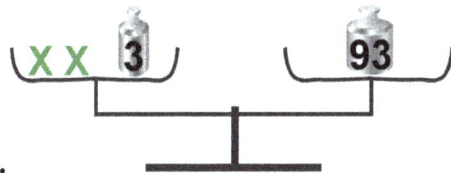

_____ = _____

_____ = _____

_____ = _____

c.

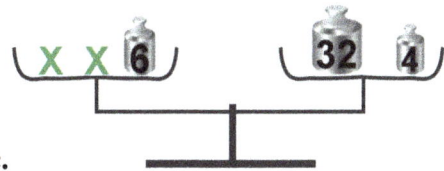

_____ = _____

_____ = _____

_____ = _____

_____ = _____

d.

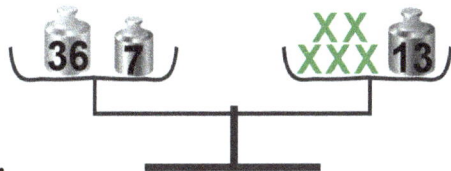

_____ = _____

_____ = _____

_____ = _____

_____ = _____

3. Solve.

a. (X 51) (X 5 X)

_____ = _____

_____ = _____

_____ = _____

_____ = _____

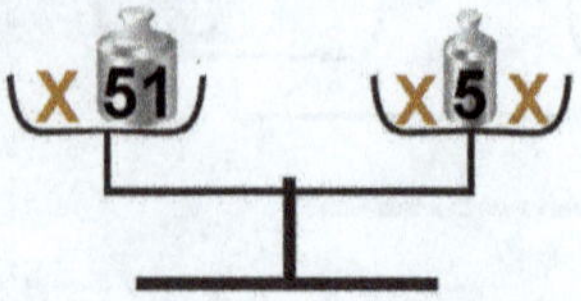

b. (9 X 6) (X X 2)

_____ = _____

_____ = _____

_____ = _____

_____ = _____

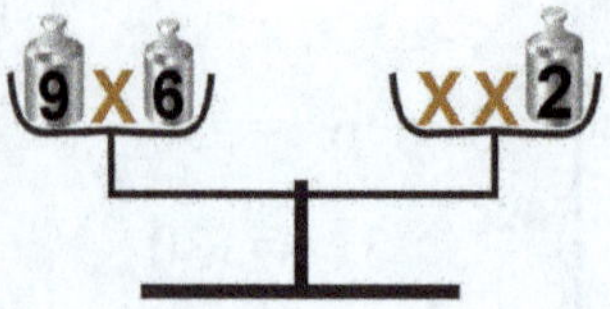

c. (XX XX 6) (X 13 5)

_____ = _____

_____ = _____

_____ = _____

_____ = _____

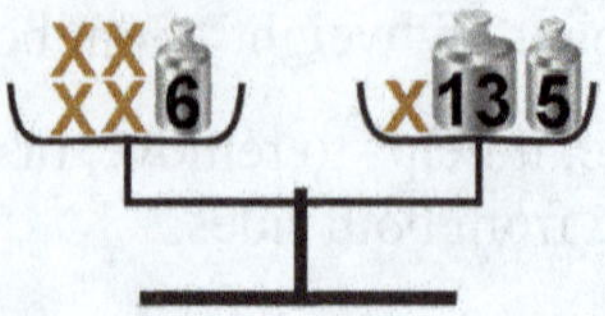

4. Solve these equations. To help you, you may draw all the "stuff" on the left and right sides on the two pans, and "remove" the same amounts from both sides to solve the equation. Or you can solve these without using the visual model. You may not need all the empty lines provided.

a.

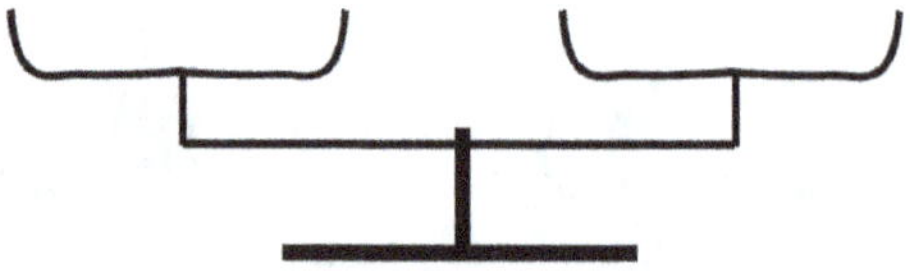

$$2x + 5 = 41$$

_____ = _____

_____ = _____

_____ = _____

b.

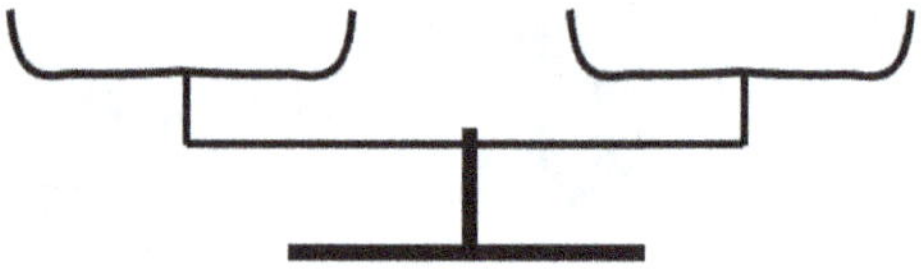

$$3x + 37 = 4x$$

_____ = _____

_____ = _____

_____ = _____

c.

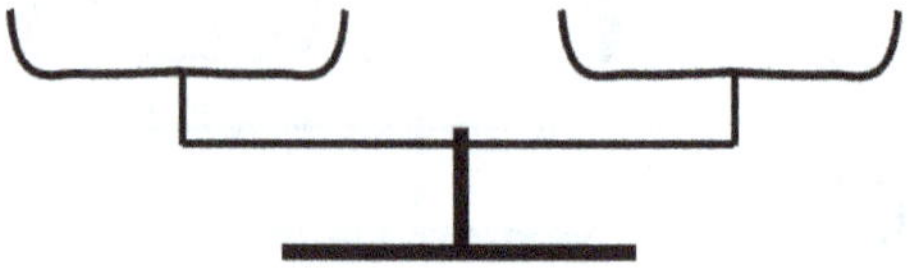

$$x + 15 = 2x + 7$$

_____ = _____

_____ = _____

_____ = _____

d.

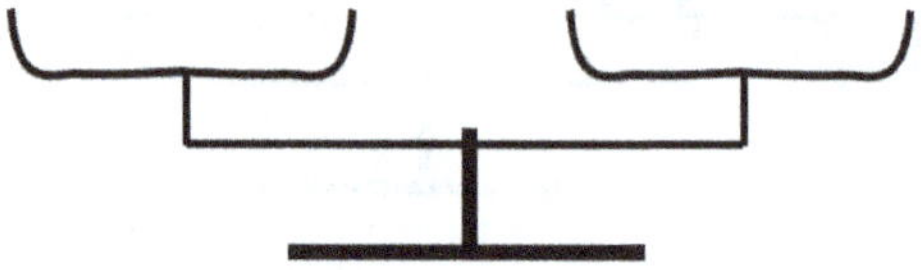

$$3x + 8 = 26$$

_____ = _____

_____ = _____

_____ = _____

Multiplying and Dividing in Parts

You have already learned about **multiplying in parts** or **partial products**. For example, you can solve 7 · 84 by multiplying 7 · 80, then multiplying 7 · 4, and then adding the two results.

Essentially, we think of the second factor 84 as the **quantity** or **sum** (80 + 4), and then multiply both of its parts separately by 7:

$$7 \cdot 84$$
$$= 7 \cdot (80 + 4)$$
$$= 7 \cdot 80 + 7 \cdot 4$$
$$= \quad 560 + 28$$
$$= \quad 588$$

We can use this idea with subtraction, also. Let's write 98 as the difference (100 − 2). We can then multiply the product 8 · 98 thinking of it as 8 · (100 − 2), and using partial products:

$$8 \cdot (100 - 2)$$
$$= 8 \cdot 100 - 8 \cdot 2$$
$$= \quad 800 - 16$$
$$= \quad 784$$

1. Write each given product using subtraction or addition. Then solve using partial products.

a. 7 · 99 = 7 · (100 − 1) = 700 − 7 = _____	**b.** 4 · 999 = 4 · (_____ − _____) =
c. 5 · 104 = 5 · (_____ + _____) =	**d.** 5 · 998
e. 6 · 98	**f.** 7 · 2,030

2. Write two expressions for the area of the whole rectangle, thinking of the large rectangle as the sum of two smaller ones. Study the example in part (a). In part (d), draw the picture yourself.

a. Total area: _3_ · (_6_ + _4_) The areas of the two rectangles: _3_ · _6_ and _3_ · _4_	**b.** Total area: _____ · (_____ + _____) The areas of the two rectangles: _____ · _____ and _____ · _____
c. Total area: _____ · (_____ + _____) The areas of the two rectangles: _____ · _____ and _____ · _____	**d.** Total area: _5_ · (_2_ + _3_) The areas of the two rectangles: _5_ · _2_ and _____ · _____

Remember **partial products** and the multiplication algorithm?

On the right, 25 · 39 is solved using partial products. The partial products are: 9 · 5, then 9 · 20, then 30 · 5, and lastly 30 · 20.

Notice there are *four* partial products. Notice also that we use 20 and 30 when we multiply, not 2 and 3. This is because the "2" in 25 really means 20, and the "3" in 39 really means 30.

```
        2 5
    x   3 9
        4 5
      1 8 0
      1 5 0
    + 6 0 0
      9 7 5
```

3. **a.** Which partial products do 80 and 700 correspond to?

```
        7 8
    x   1 6
        4 8
      4 2 0
       (8 0)
    + (7 0 0)
    1 2 4 8
```

b. Solve using partial products.

```
        5 6
    x   8 4
```

c. Solve using partial products.

```
        1 7
    x   9 5
```

Example 1. The picture illustrates the multiplication 38 · 57 using an area model. Study it carefully. It corresponds *exactly* to the partial products algorithm above: the total area is solved *in parts*. The total area of the rectangle is:

$$38 \cdot 57 = \; 30 \cdot 50 + 30 \cdot 7$$
$$+ \; 8 \cdot 50 + 8 \cdot 7$$
$$= \; 1{,}500 + 210 + 400 + 56 = 2{,}166 \textbf{ square units}$$

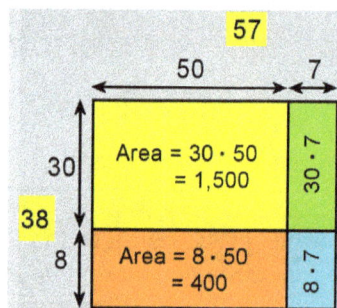

4. The rectangular area models illustrate two multiplications (not to scale). In each rectangular part, write how many square units its area is. Then, find the total area by adding the areas of the parts.

a. 29 · 17

b. 75 · 36

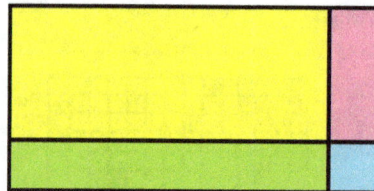

We can also **divide a sum or difference in parts**.

Example 2. In the quotient $\dfrac{40+55}{5}$, we can divide $\dfrac{40}{5}$ and $\dfrac{55}{5}$ separately, and then add the results.

We get $\dfrac{40}{5} + \dfrac{55}{5} = 8 + 11 = 19$.

Example 3. Dividing in parts works equally well with subtraction: $\dfrac{120-48}{4} = \dfrac{120}{4} - \dfrac{48}{4} = 30 - 12 = 18$.

5. Divide in parts, then add or subtract the results.

a. $\dfrac{80+12}{2}$	**b.** $\dfrac{350+15}{5}$	**c.** $\dfrac{400-12}{4}$
d. $\dfrac{9{,}300-60}{3}$	**e.** $\dfrac{350+21-7}{7}$	**f.** $\dfrac{900-18}{9}$
g. $\dfrac{22 \text{ m } 9 \text{ cm}}{2}$	**h.** $\dfrac{40 \text{ kg } 750 \text{ g}}{5}$	**i.** $\dfrac{12 \text{ L } 600 \text{ ml}}{4}$

How can you make sense of this? Let's say you have both apples and oranges in a bag, and you are going to share them equally between 5 people. How many pieces of fruit will each person get? You could just mix all the pieces of fruit and divide the total number by five to find the answer, but you can also take only the apples and divide those by 5, and then take only the oranges and divide them by five. In essence:

$$\frac{\text{apples} + \text{oranges}}{5} = \frac{\text{apples}}{5} + \frac{\text{oranges}}{5}$$

(Of course, you probably want to divide the fruit separately in this situation, and not mix them. But the NUMBER of pieces of fruit that each person would get can be found either way.)

6. Divide in parts in your head. First, think how the dividend can be written in two or more parts.

a. $\dfrac{412}{2}$	**b.** $\dfrac{609}{3}$	**c.** $\dfrac{824}{8}$	**d.** $\dfrac{1{,}206}{6}$	**e.** $\dfrac{4{,}518}{9}$

7. You have 2 liters 250 milliliters of ice cream that you want to share equally with three of your friends (four people in total) at a birthday party. How much ice cream will each person get?
Round your answer to the nearest 10 milliliters.

Dividing $\dfrac{21+2}{7}$ in parts, we get $\dfrac{21}{7}$ and $\dfrac{2}{7}$. While 21/7 is just 3, the other part, 2/7, has to be left as a fraction. We get $3 + \dfrac{2}{7} = 3\dfrac{2}{7}$. Of course, this is the same as writing the fraction $\dfrac{23}{7}$ as a mixed number.

8. Divide in parts. You will have a fraction in the answer.

a. $\dfrac{15+4}{5}$	**b.** $\dfrac{44+7}{11}$	**c.** $\dfrac{6+70}{7}$
d. $\dfrac{420+2}{6}$	**e.** $\dfrac{240+12+3}{4}$	**f.** $\dfrac{2+36+270}{9}$

9. Divide in parts in your head. First, think how the dividend can be written in two or more parts. See the example.

a. $\dfrac{403}{4} = \dfrac{400+3}{4} =$	**b.** $\dfrac{911}{3}$
c. $\dfrac{5{,}024}{5}$	**d.** $\dfrac{81}{4}$
e. $\dfrac{127}{3}$	**f.** $\dfrac{365}{6}$

Reminder: you can *only* divide in parts when there is a single number in the denominator (the divisor). In the expression $\dfrac{30+120}{3+7}$, we need to first solve $3+7$. After that, you could divide in parts.

You could also simply calculate the two sums first to get 150/10 = 15.

10. Simplify. In some of these problems, it helps to divide in parts. Can you find which ones?

a. $\dfrac{3+4}{5+9}$	**b.** $\dfrac{12-5}{3+13+5}$
c. $\dfrac{30+50}{2+9}$	**d.** $\dfrac{6+24+240}{8}$
e. $\dfrac{120-3}{7-3}$	**f.** $\dfrac{100}{80-50}$

Fill in the blanks so the equations are true.

Puzzle Corner

a. $\dfrac{\boxed{} - \boxed{}}{10} = 25 - \dfrac{3}{10}$

b. $\dfrac{\boxed{} - 3}{5} = 2\dfrac{1}{5} - \dfrac{\boxed{}}{\boxed{}}$

More Mental Math

To **multiply** 2,000 × 120, simply multiply 2 × 12, and place four zeros on the end of the answer: $\underline{2,000} \times \underline{120} = \underline{24}0,000$	Solve **division** by thinking of multiplication "backwards": $5,600 \div 70 = ?$ Think what number times 70 will give you 5,600. Since 70 × 80 = 5,600, then 5,600 ÷ 70 = 80.	You can **add in parts**. $76 + 120 + 65 = ?$ First add 70 + 120 + 60 = 250. Then, 6 + 5 = 11. Lastly, 250 + 11 = 261.

The **order of operations** is:
1. Parentheses 2. Exponents; 3. Multiplication and division; 4. Addition and subtraction.

To calculate 9 × 80 − 10 × 70, first solve 9 × 80 and 10 × 70 . Subtract only after those calculations. $9 \times 80 - 10 \times 70$ $= 720 - 700 = 20$	In the expression 4,500 ÷ (5 + 45) × 80, solve 5 + 45 first. Then, divide. $4,500 \div (5 + 45) \times 80$ $= 4,500 \div 50 \times 80$ $= 90 \times 80 = 7,200$

1. Solve in your head.

a. $410 + 2 \times 19$ =	**b.** $3 \times 50 + 4 \times 150$ =	**c.** $70 \times 80 - 40 \times 50$ =
d. $14 + (530 - 440)$ =	**e.** $45 + 56 + 35$ =	**f.** $300 \div 5 - 400 \div 10$ =

2. Solve in your head.

a. $17 + \underline{\hspace{1.5cm}} = 110$ **b.** $345 + \underline{\hspace{2cm}} = 1,000$ **c.** $3 \times 40 + \underline{\hspace{2cm}} = 500$

3. Divide. Remember that division can also be written using a fraction line.

a. $\dfrac{240}{4} =$ **c.** $\dfrac{72}{9} =$ **e.** $\dfrac{5,600}{10} =$ **g.** $\dfrac{420}{20} =$ **i.** $\dfrac{420}{70} =$

b. $\dfrac{7,200}{100} =$ **d.** $\dfrac{450}{9} =$ **f.** $\dfrac{8,000}{200} =$ **h.** $\dfrac{10,000}{50} =$ **j.** $\dfrac{7,200}{800} =$

4. Solve. Notice carefully which operation(s) are done first.

a. $500 - 40 - 3 \times 50 =$ _____	**b.** $1,020 - (40 - 10) \times 20 =$ _____
c. $42,000 - 12,000 + 3 \times 5,000 =$ _____	**d.** $(70 - 20) \times 70 =$ _____
e. $\dfrac{210}{2} + 3 \times 15 =$ _____	**f.** $250 \times 4 + \dfrac{6,300}{70} =$ _____

5. Find a number that fits in place of the unknown.

a. $x \div 70 = 40$	**b.** $20 \times M = 1,200$	**c.** $500 - y = 320$

6. Find the rule that is used in the table and fill in the missing numbers.

n	130	250	360	410	775	820	1,000
$n -$ _____		215		375			

7. Find the rule that is used in the table and fill in the missing numbers.

n	3	5	12	15	25	35	60
		200			1,000		

8. Rick cut off a 50-cm piece from a 6-meter board, and then he divided the rest of the board into five equal pieces. How long was each piece?

9. **a.** Evelyn works 8 hours a day and earns $104 daily. What is her hourly wage?

   b. How much does Evelyn earn in a five-day work week?

   How much does she earn in three months (which is 13 weeks)?
   (You may use paper and pencil for this one.)

10. Alexis and Mia baked biscuits for a bake sale. They used this recipe, but they needed to triple it:

    a. Triple the recipe for them.

    b. How many biscuits did they bake?

    2 1/4 cups of flour
    3 teaspoons of baking powder
    1/3 cup of honey
    1/2 cup of butter
    3/4 teaspoon of nutmeg
    1 1/2 teaspoons of cinnamon
    1/2 teaspoon of ground cloves
    3/4 cup of walnuts
    Makes 2 1/2 dozen biscuits.

Review of the Four Operations 1

1. Use the following problems to review long division and multiplication.

a.
```
        5 3 6
    x     7 1
```

b.
```
    $ 2 4.5 9
    x       7 0
```

c.
```
        2 0 6
    x   9 1 5
```

d.
```
    7)2 0 5 8
```

e.
```
    7)5 9 9.2
```

f.
```
    8)8 2 7 2
```

2. How do you check the result of *any* division problem? *(Hint: check the next page.)*

Now, check your answers for 1. d, 1. e and 1. f.

Long division example, four stages:

$$
\begin{array}{r}
1\,3 \\
6\,\overline{)\,8\,0\,1\,2\,9\,8} \\
-6 \\
\hline
2\,0 \\
-1\,8 \\
\hline
2
\end{array}
\qquad
\begin{array}{r}
1\,3\,3 \\
6\,\overline{)\,8\,0\,1\,2\,9\,8} \\
-6 \\
\hline
2\,0 \\
-1\,8 \\
\hline
2\,1 \\
-1\,8 \\
\hline
3
\end{array}
\qquad
\begin{array}{r}
1\,3\,3\,5 \\
6\,\overline{)\,8\,0\,1\,2\,9\,8} \\
-6 \\
\hline
2\,0 \\
-1\,8 \\
\hline
2\,1 \\
-1\,8 \\
\hline
3\,2 \\
-3\,0 \\
\hline
2\,9
\end{array}
\qquad
\begin{array}{r}
1\,3\,3\,5\,4\,9 \\
6\,\overline{)\,8\,0\,1\,2\,9\,8} \\
-6 \\
\hline
2\,0 \\
-1\,8 \\
\hline
2\,1 \\
-1\,8 \\
\hline
3\,2 \\
-3\,0 \\
\hline
2\,9 \\
-2\,4 \\
\hline
5\,8 \\
-5\,4 \\
\hline
4
\end{array}
$$

Long division works the same way when there are several digits in the dividend (the big number we divide into). Study the example carefully.

The answer we get is 801,298 ÷ 6 = 133,549 R4.

3. Divide using long division.

a.
$$ 7\,\overline{)\,4\,2\,3\,3\,6} $$

b.
$$ 6\,\overline{)\,2\,0\,9.7\,0} $$

c.
$$ 5\,\overline{)\,5\,4\,9\,2\,0\,7} $$

To check a division result that has a remainder, multiply the result by the divisor, and then *add* the remainder. You should get the original dividend.

In this case, we multiply and add: 6 × 133,549 + 4 = 801,298, so it checks.

Remember that the remainder is always less than the divisor; if it isn't, you can continue the division!

4. Check each division by multiplying and adding. If the division is incorrect, correct it.

a. 437 ÷ 6 = 72 R5	b. 2,045 ÷ 3 = 681 R1
_____ × _____ + _____ =	_____ × _____ + _____ =

5. A bakery bagged 177 buns into bags of eight, getting 21 bags, and nine buns left over. The division was: $177 \div 8 = 21$ R9. Jessica *immediately* spotted this was wrong (without calculating anything). How did she do that?

6. A large school has 542 sixth graders. How would you divide them into classes as evenly as possible, with about 25 students per class?

7. Divide, using two-digit divisors. You can build a multiplication table for the divisor to help you. Lastly, check your result.

$2 \times 45 = 90$	**a.** $45 \overline{)4\ 0\ 0\ 5}$	$\times\quad 4\ \ 5$
$2 \times 75 = 150$	**b.** $75 \overline{)1\ 9\ .\ 8\ 7\ 5}$	$\times\quad 7\ \ 5$

33

8. Divide, using two-digit divisors. These may have a remainder. You can build a multiplication table for the divisor to help you. Lastly, check your result.

$2 \times 48 = 96$	**a.** $48 \overline{)8\ 7\ 0\ 2\ 5}$	$\times\ \ 4\ \ 8$
$2 \times 90 = 180$	**b.** $90 \overline{)8\ 7\ 1\ 6\ 6\ 0}$	$\times\ \ 9\ \ 0$
$2 \times 82 = 164$	**c.** $82 \overline{)5\ 4\ 0\ 2\ 2}$	$\times\ \ 8\ \ 2$

34

9. Try your division skills with 3-digit divisors, too. The answer key has the complete solution, if you get "stuck."

	a. $101\overline{)299046}$	$\begin{array}{r} \times\ 1\ 0\ 1 \\ \hline \end{array}$
	b. $123\overline{)3634206}$	$\begin{array}{r} \times\ 1\ 2\ 3 \\ \hline \end{array}$
	c. $350\overline{)7652000}$	$\begin{array}{r} \times\ 3\ 5\ 0 \\ \hline \end{array}$

35

10. Here are some riddles for you to solve for more practice with long division! Use your notebook.

I	$42{,}408 \div 76$	E	$44{,}217 \div 51$	E	$128{,}316 \div 111$
M	$85{,}104 \div 54$	I	$223{,}496 \div 91$	E	$51{,}313 \div 97$
O	$23{,}530 \div 26$	I	$30{,}624 \div 33$	M	$880{,}341 \div 309$
R	$61{,}880 \div 35$	R	$133{,}140 \div 70$	T	$113{,}168 \div 88$
V	$51{,}944 \div 86$	S	$11{,}880 \div 22$	R	$693{,}360 \div 810$

What is as round as a dishpan, and no matter the size, all the water in the ocean cannot fill it up?

540   558   529   604   1,156

What flies without wings?

1,286   928   1,576   867

I am the only thing that always tells the truth. I show off everything that I see.

2,849   2,456   1,768   1,902   905   856

G	$200{,}196 \div 201$	R	$617{,}105 \div 415$	O	$1{,}388{,}740 \div 230$
O	$324{,}729 \div 57$	S	$2{,}863{,}250 \div 250$	P	$759{,}290 \div 70$
E	$339{,}388 \div 31$	T	$1{,}049{,}664 \div 88$	I	$678{,}040 \div 506$
S	$2{,}337{,}820 \div 205$	H	$236{,}215 \div 35$	T	$250{,}536 \div 44$
E	$28{,}548 \div 18$	F	$97{,}920 \div 16$	F	$239{,}397 \div 199$

From what heavy seven-letter word can you take away two letters and have eight left?

1,203   1,487   1,586   1,340   996   6,749   5,694

The more of them you take, the more you leave behind. What are they?

6,120   6,038   5,697   11,928   11,453   11,928   10,948   10,847   11,404

36

Review of the Four Operations 2

1. Last year, in the Gordon family, Father earned $29,600, Mother earned $13,500, and Matt earned $8,300. They figured out that they had paid about 1/5 of their total earnings in taxes, and used about 1/4 of their income for groceries.

   a. Calculate how much the family used for groceries.

   b. What fractional part of their income did the family have left to spend, after taxes and groceries?

2. Find the value of these expressions, using paper and pencil methods. Use your notebook for more space.

   a. $100 - 29.5 \times 2.6$

   b. $2.3 + 9.356 + 0.403 + 908.8$

   c. $800 - (12.48 - 2.9)$

   d. $559.50 \div 3$

3. Write the division equation, if the calculation to check it is $13 \times 381 + 5 = 4{,}958$.

4. **a.** If you need to solve $65 \div 7$ to three decimal digits using long division, how many decimal zeros should you add to 65 before starting the division?

   b. Solve $65 \div 7$ to three decimal digits.

5. A large gym floor measures 10 m by 12 m. The teacher divides that into nine equal-sized areas. How big is each area in square meters? Give your answer to two decimal digits.

6. An apple harvest produced 2,350 kg of apples. The farmer packed 36 apples per box. One apple weighs approximately 250 grams. How many boxes were needed to pack the apples?

7. **a.** A car is traveling at 54 miles per hour. Fill in the table:

Miles				54 Miles			
Time	10 min	20 min	30 min	1 hour	2 hours	2 1/2 hours	3 hours

b. If the Jones family travels steadily at 54 miles per hour, how far will they travel in 9 hours?

c. *Estimate* how many *hours* it takes them to travel 550 miles.

8. Dad drives at a constant speed of 40 miles per hour.

a. How many minutes does it take him to travel 5 miles?

b. How about 100 miles?

c. Dad drives 30 miles to work. What time should he leave to arrive at exactly 9:00 am?

9. A company bought 96 gallons of fruit juice for a total of $3,072. They packaged it into 8-ounce bottles.

a. How many jars did they fill?

> *Think: how many ounces are in a gallon?*

b. What is the minimum price that they would have to charge per bottle to get back at least what they paid ("break even")?

Puzzle Corner   Find what is missing from the equations. You do *not* need to calculate anything!

a. $4,392 - \underline{\hspace{1cm}} + 293 = 4,392$ **b.** $384 \div 8 \times \underline{\hspace{1cm}} = 384$ **c.** $\dfrac{1,568}{49} \times \underline{\hspace{1cm}} = 1,568$

Lessons in Problem Solving

You can draw a **flowchart** to help you find the logical way to solve a multi-step problem.

Write in the flowchart *what* you plan to solve in each step. You can also write down other notes you feel are important about that step.

Note 1: You don't have to use a flowchart. Its purpose is to help you organize your thoughts. By making it, you are essentially finding the solution path—which is the most important and more challenging part of problem solving. After that, all you have to do is the calculations, which is the easier, mechanical part.

Note 2: Finding the "path" to the solution can take some time, and perhaps you will find that you need to backtrack a step and find another way. This is normal. Even in real life, problems in all areas of life are solved with a process that includes several attempts, backtracking, and changing plans. Think of problem solving as a <u>process</u>—not as something you "have" to get right the first time.

Look at the examples below, and solve the problems. Use extra paper for calculations if needed.

---

1. A large carpet costs $55.50, and a small one costs 2/5 of that price. Luis bought two of the smaller carpets. What was his change from $50?

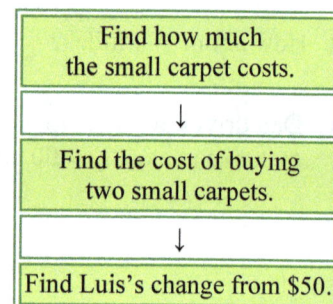

> Find how much the small carpet costs.
> ↓
> Find the cost of buying two small carpets.
> ↓
> Find Luis's change from $50.

---

2. Angela has two kinds of plastic containers. The larger ones hold 0.75 liters, and the smaller ones hold 7/10 of that amount.

   Can Angela fit 5 liters of soup into four large and five small containers?

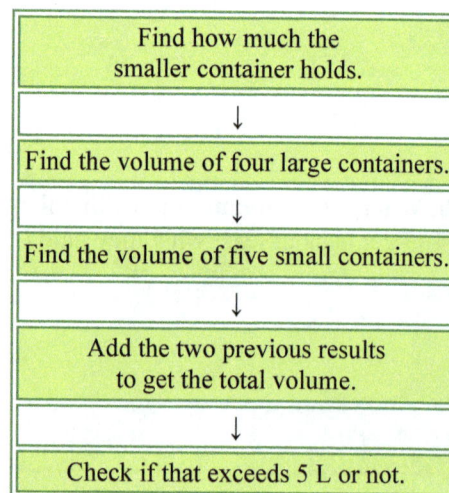

> Find how much the smaller container holds.
> ↓
> Find the volume of four large containers.
> ↓
> Find the volume of five small containers.
> ↓
> Add the two previous results to get the total volume.
> ↓
> Check if that exceeds 5 L or not.

3. A 25-kg box of bolts was divided equally into 20 bags, and similarly,
a 15-kg box of nuts was divided equally into 20 other bags.
How much would one bag of bolts and one bag of nuts weigh together?

Obviously you need to divide. Look at the three different ways to do the first division:

Way 1:	Way 2:	Way 3:
Divide 25,000 grams by 20. Your answer will be in grams.	Divide 25.000 kg by 20. Your answer will be a decimal and in kilograms.	Divide first by 10, and then the result by 2.

Here is a "flowchart" to illustrate the solution process.

Now you solve the problem.

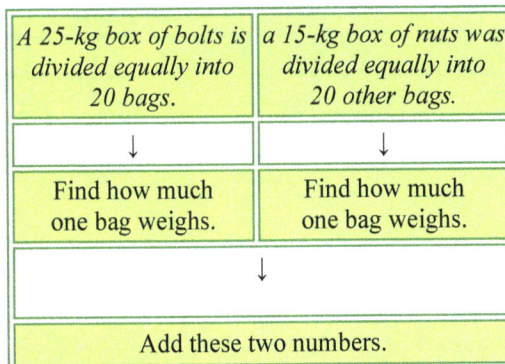

A 25-kg box of bolts is divided equally into 20 bags.	a 15-kg box of nuts was divided equally into 20 other bags.
↓	↓
Find how much one bag weighs.	Find how much one bag weighs.
↓	
Add these two numbers.	

4. A company sells jars of jam in three different sizes. The largest size is 670 g,
the medium size is 3/4 of that, and the smallest size is 2/3 of the medium size.

a. Find the weight of the medium and smallest-sized jars.
Round the weights to the nearest gram.

b. Find the total weight of one large, one medium, and one small-sized jar.

5. John spent 4/9 of his money and Karen spent 4/7 of hers. Now they each have $30.60 left. How much more did Karen have initially than John?

You need to read this carefully and solve it in parts. To find out *how much more*, we need to know both numbers. So, first we need to find out how much Karen had and how much John had initially. Both of those can be solved separately using the bar model method.

The bar model below will help you solve how much money John had initially.

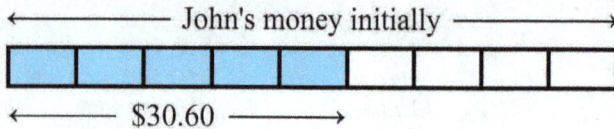

←——————— John's money initially ———————→

←——— $30.60 ———→

John spent 4/9 of his and had $30.60 left.	Karen spent 4/7 of hers and had $30.60 left.
↓	↓
Find how much John had initially.	Find how much Karen had initially.
↓	
Subtract these two numbers to find the difference.	

6. A company is taking 569 employees to a water park. It is 30 miles each way.
   Each bus seats 43 people. The cost for each bus is $2.15 per mile.

   a. How many buses do they need?

   b. What are the transportation costs?

7. A washing machine has been discounted by 1/10 of its price, and now it costs $360.
   Another washer has been discounted by 2/5 of its price, and now it costs $348.
   Find the price difference between the two washers *before* the discount.
   Hint: Draw two bar models, one for the price of each washer.

Puzzle Corner **a.** The decimal point key does not work in Henry's calculator.
But Henry discovered a way to enter decimals into his
calculator without using the decimal point key!

Find how he did it for these numbers: 0.1, 0.81, 0.492 and 3.55.

b. Find a way to calculate 1.38 × 0.39 with the calculator, without using the decimal point key.

The Coordinate Grid

This is a **coordinate grid**. It consists of two number lines that are set perpendicular (at right angles) to each other.

The horizontal number line is called the **x-axis**. The vertical one is called the **y-axis**.

You can see one point, called "A," that is drawn or *plotted* on the grid.

Since we have two number lines, we use *two* numbers (4 and 6) to signify its location. Those numbers are the **coordinates** of the point A.

The first number, 4, is the **x-coordinate** of the point A. It is called the *x*-coordinate because point A is <u>four units from zero</u> in the horizontal direction (direction of the *x*-axis).

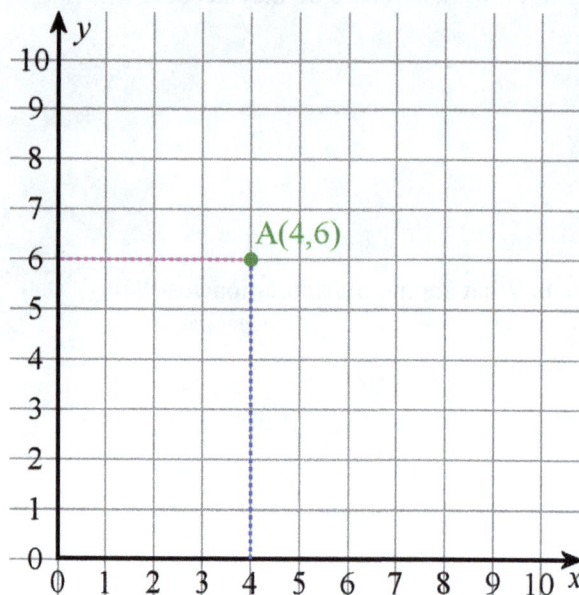

We can see that by drawing a straight line down from A. The line *intersects*, or "hits," the *x*-axis at 4.

The second number is the **y-coordinate** of the point A. <u>In the vertical direction, point A is six units from zero</u>. When we draw a line directly towards left from A, it intersects the *y*-axis at 6.

We write the two coordinates of a point inside parentheses, separated by a comma: (4, 6).

Note: (4, 6) is an **ordered pair**: the order of the two coordinates matters. The *first* number is ALWAYS the *x*-coordinate, and the *second* number is always the *y*-coordinate, not vice versa.

1. Write the two coordinates of the points plotted on the coordinate grid. For points A and B, the helping lines are drawn in. (The helping lines are not necessary to draw; they are just that — *helping* lines. You can draw them if they help you.)

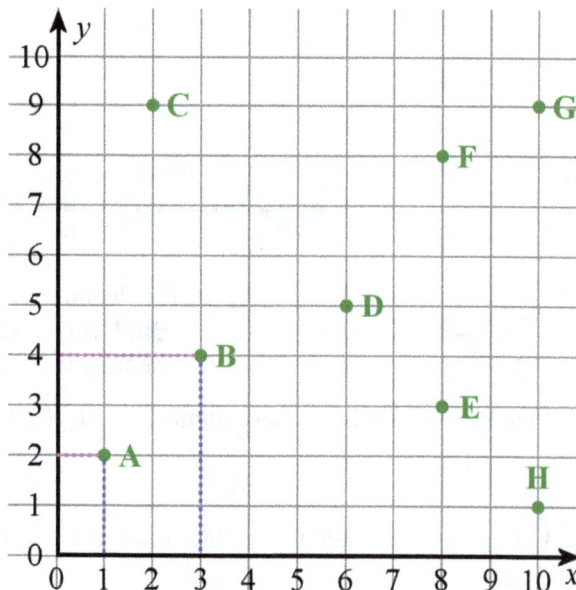

A (___ , ___) B (___ , ___)

C (___ , ___) D (___ , ___)

E (___ , ___) F (___ , ___)

G (___ , ___) H (___ , ___)

44

To plot points, you can first "travel" on the *x*-axis from the point (0, 0) (the **origin**) the number of units indicated by the *x*-coordinate.

Then travel UP as many units as the *y*-coordinate indicates.

The image shows an example of how to plot (7, 5).

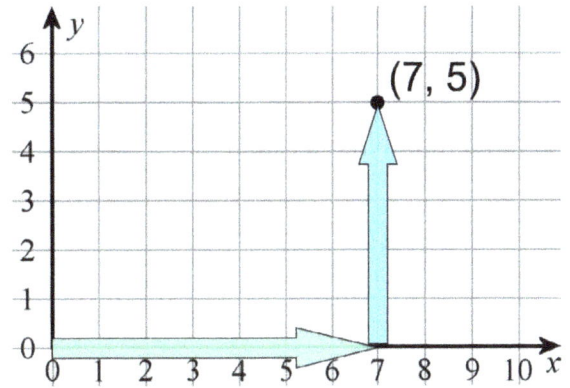

2. Plot the following points on the coordinate grid. Then join them with line segments in the alphabetical order. What do you get?

A(1, 5)   B(4, 3)   C(4, 6)

D(7, 5)   E(6, 8)

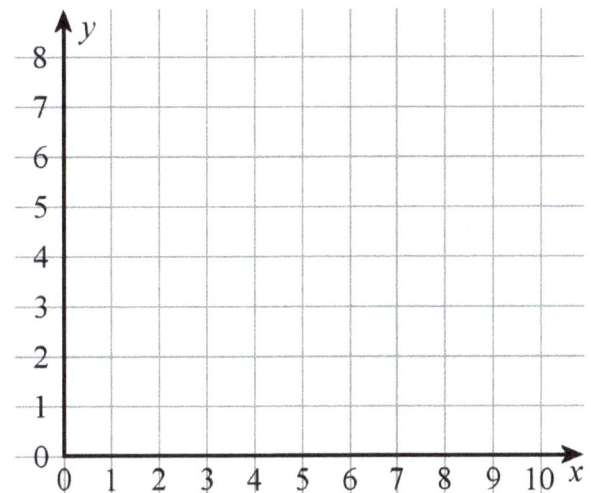

3. **Zero as a coordinate.** Plot the following points in the grid on the right.

A(0, 6)   B(0, 3)   C (0, 0)

D(5, 0)   E(9, 0)

What do you notice?

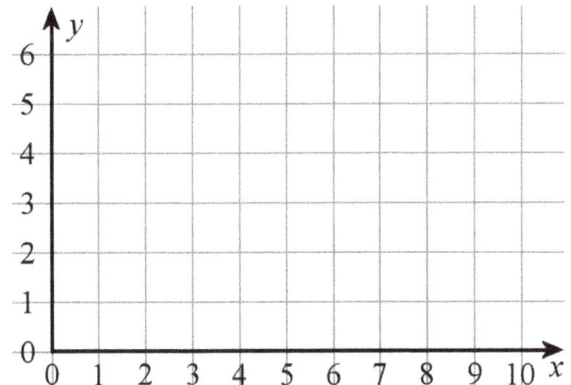

4. **a.** Write the coordinates of the points E, F, and G.

   b. Plot a fourth point, H, so that when you join E, F, G, and H with line segments, you will get a rectangle.

   c. What are the coordinates of H?

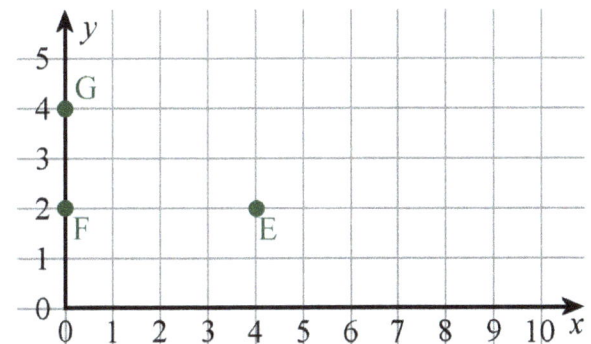

45

5. In this grid, the *y*-axis is scaled differently.

   a. Write the coordinates of these points:

   A(____ , ____)   B(____ , ____)

   C(____ , ____)

   b. Plot these points. Note that the points don't necessarily fall on the gridlines.

   D(7, 11)   E(1 ½, 9)   F(9 ¼, 2)

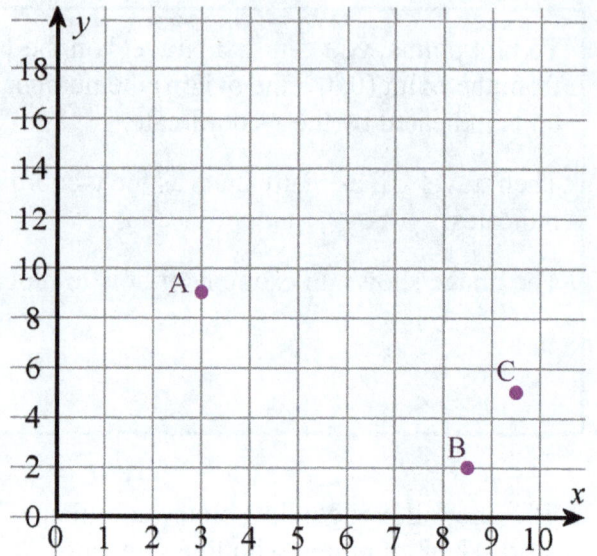

6. **a.** Design a scaling for the axes so that the point P(36, 38) will fit on this grid.

   b. Then plot these points also, and connect the points with line segments in order. What shape is formed?

   Q(36, 28)   R(16, 18)   S(26, 38)

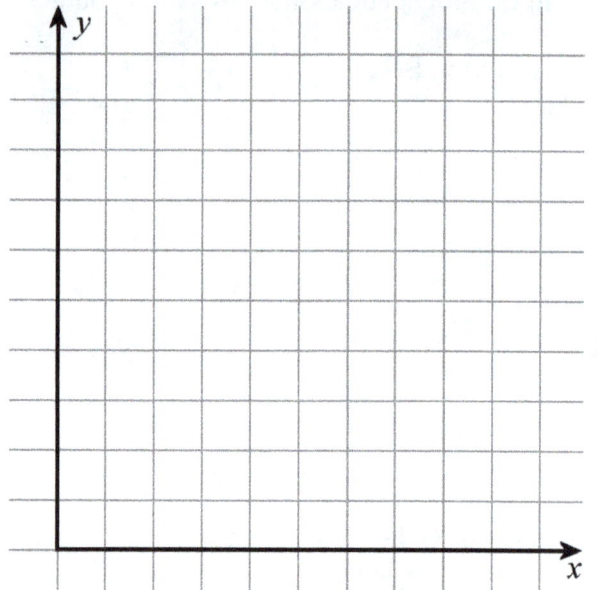

7. Here, "LINE (5,6) - (2,7)" means a line segment that is drawn from (5, 6) to (2, 7).

   Draw the following line segments (joining the two given points). Use a ruler! The first one is already done for you.

   What figure is formed?

   LINE (1, 0) - (1, 5)   LINE (1, 5) - (0, 5)

   LINE (0, 5) - (4, 7)   LINE (4, 7) - (8, 5)

   LINE (8, 5) - (7, 5)   LINE (3, 0) - (3, 3)

   LINE (5, 0) - (5, 3)   LINE (3, 3) - (5, 3)

   LINE (1, 0) - (7, 0)   LINE (7, 0) - (7, 5)

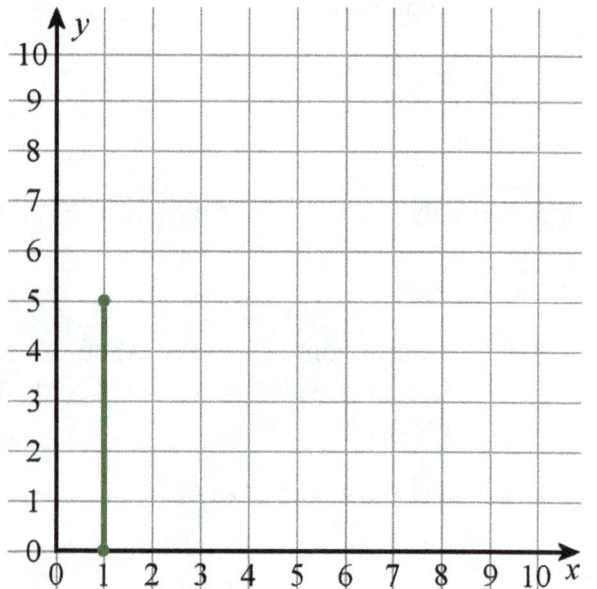

46

The Coordinate Grid, Part 2

1. Plot the points (3, 2) and (3, 7), and draw a vertical line through them.

   Think carefully. Which of the following points would also be on this line, if you extended it beyond the grid?

   (11, 3)   (3, 75)   (23, 7)   (3, 37)

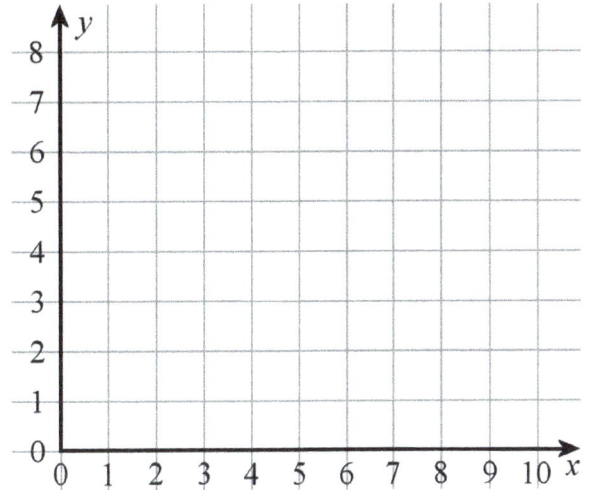

2. Plot the points (2, 4) and (8, 4), and draw a horizontal line through them.

   Think carefully. Which of the following points would also be on this line, if you extended it beyond the grid?

   (18, 2)   (19, 8)   (35, 4)   (4, 19)

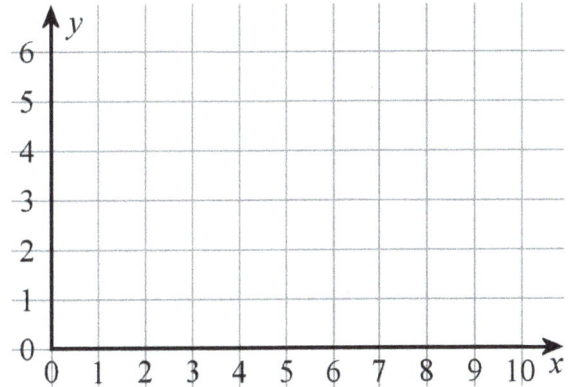

3. Liz drew a vertical line through (2, 8).
   What is true of the coordinates of *every* point on her line?

4. In this exercise you will draw triangles.

   Each set of three points below defines a triangle. Plot them. (Color them if you'd like!)

   Triangle 1: (3, 3), (3, 5), (5, 4)

   Triangle 2: (4, 2), (6, 2), (5, 4)

   Triangle 3: (4, 6), (6, 6), (5, 4)

   Now draw a fourth triangle that also has one vertex at (5, 4) to complete the pattern.

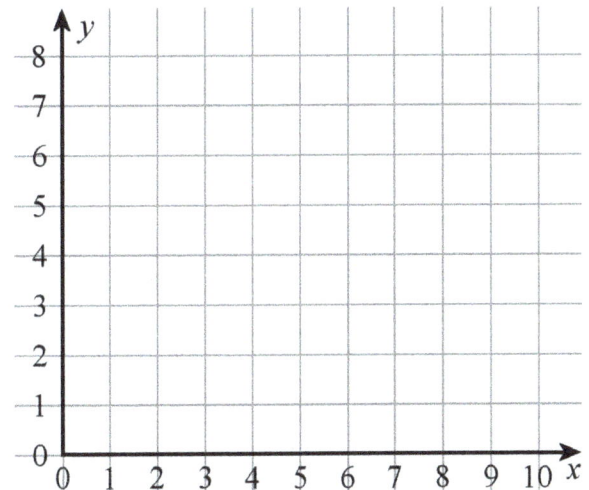

5. **a.** Design a scaling for the axes so that the point (75, 90) will fit on the grid.

   b. The points (75, 90), (75, 50), and (40, 90) are three vertices of a rectangle.

   Figure out the coordinates of the fourth vertex, and plot the rectangle.

   c. What is the area of the rectangle?

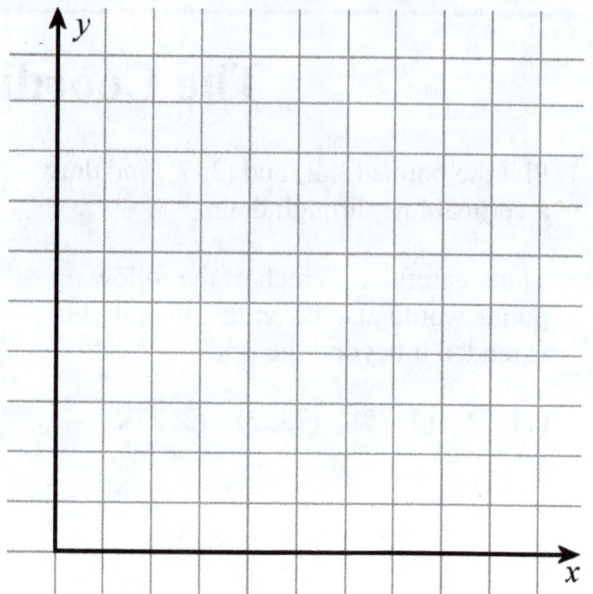

6. Create a symmetrical figure by reflecting the points across the line.

   Point C is already reflected, to become the point C' (C prime). Notice that C' is at the same distance from the line as C (two units).

   (The points already on the line do not change.)

   Lastly, join the points with line segments to get a symmetrical figure.

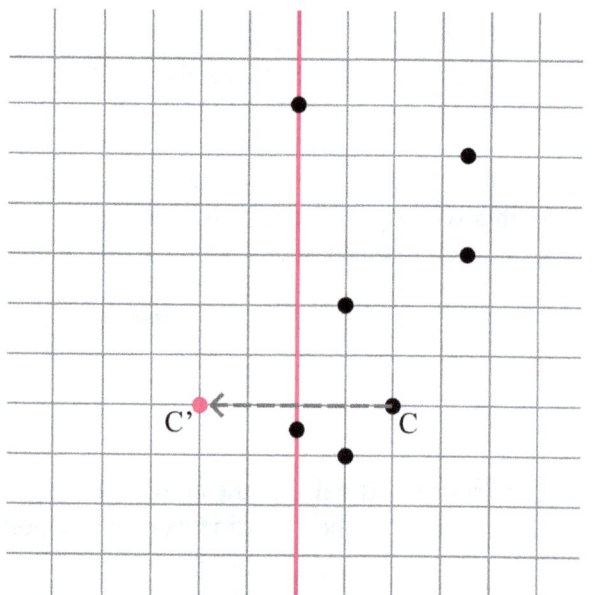

7. Design your own symmetrical design!

   Make sure that any vertices are carefully reflected across the line, so that the point and its reflected point are at the same distance from the line.

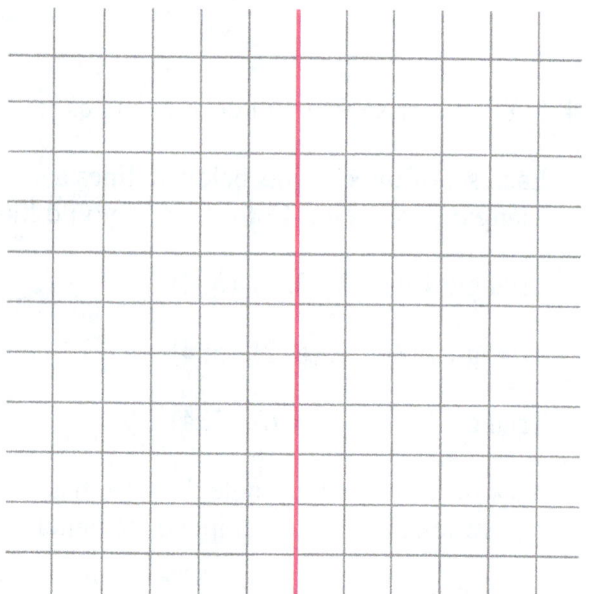

Number Patterns in the Coordinate Grid

Example 1. Look at this table. What do you notice?

x	1	2	3	4
y	2	3	4	5

The x-values (the top row) is a very simple pattern created from the rule: **Start at 1, and add 1 each time.**

The y-values (the bottom row) come from an equally simple rule: **Start at 2, and add 1 each time**.

We can look at each *column* as a number pair. These number pairs (1, 2), (2, 3), (3, 4), and (4, 5) are <u>four points</u> on the coordinate grid (see the image).

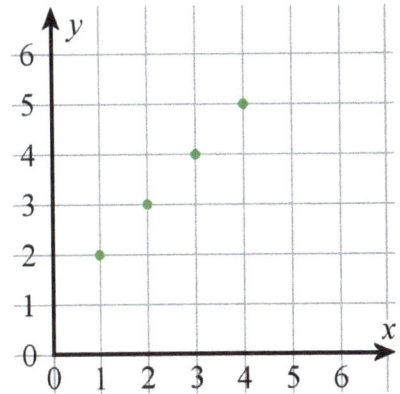

Lastly, if we look at the number pairs (1, 2), (2, 3), (3, 4), and (4, 5), we can see there is a <u>simple connection</u> or relationship between each x and y coordinate. This relationship, or rule, is: each time, **y is 1 more than x**. That rule is true for *each* of the four points.

We can also write this with symbols: **$y = x + 1$**.

1. **a.** Fill in the x and y values according to the given rules.

    <u>The rule for x-values</u>: start at 0, and add 1 each time.

    <u>The rule for y-values</u>: start at 0, and add 2 each time.

x	0	1				
y	0	2				

   b. Plot the points formed by the number pairs.

   c. What simple relationship exists between each x and y coordinate?

   d. Why do you think this relationship is there? (Where does it stem from?)

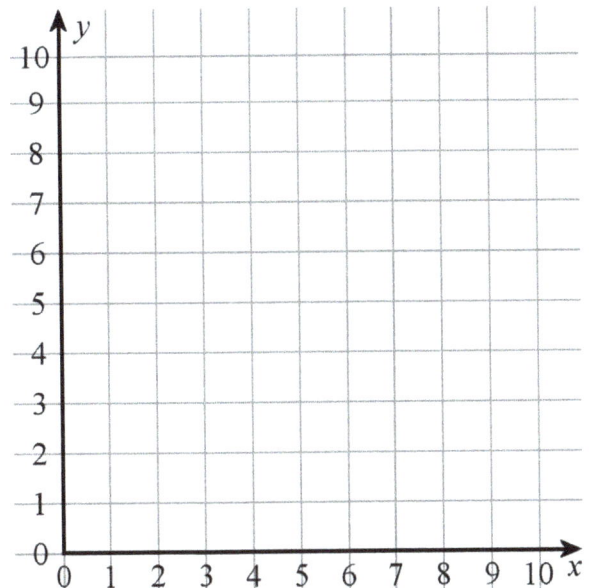

Example 1.

The rule for *x*-values:
start at 0, and add 3 each time.

The rule for *y*-values:
start at 0, and add 1 each time.

x	0	3	6	9	12	15
y	0	1	2	3	4	5

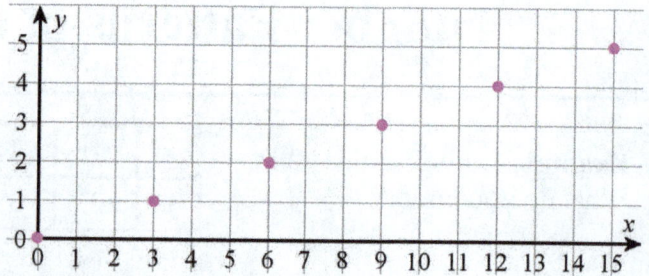

Notice that in each case, the *y*-coordinate is 1/3 of the *x*-coordinate! Or, the *x*-coordinate is three

times the *y*-coordinate. We can write this as an equation: $y = \dfrac{x}{3}$ or $x = 3y$. (Note: 3*y* means 3 times *y*.)

Why is that? Because when one variable counts by ones and the other counts by 3s, the relationship between them naturally has to do with multiplication or division by 3.

In questions 2-3, fill in the *x* and *y* values according to the given rules. Then plot the points.

2. **a.** The rule for *x*-values: start at 0, and add 2 each time.

The rule for *y*-values: start at 0, and add 1 each time.

x	0	2	4			
y	0	1	2			

b. What simple rule ties the *x* and *y*-coordinates together in each case?

c. Why is this relationship there? (Where does it stem from?)

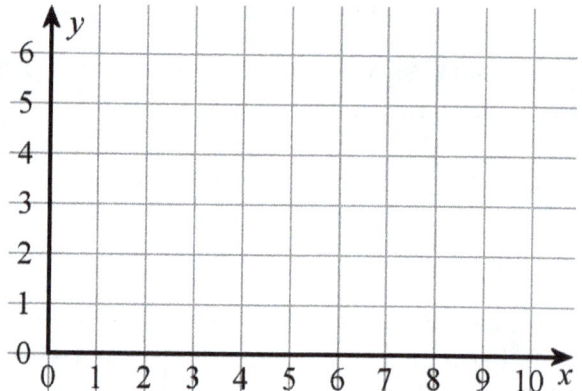

3. **a.** *x*-values: start at 0, and add 1 each time.

y-values: start at 6, and subtract 1 each time.

x						
y						

b. What simple relationship exists between each *x* and *y* coordinate?

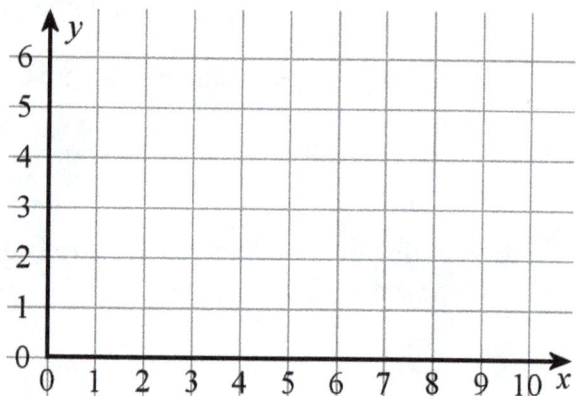

4. Write the number pairs in the table, using the plot. Then, fill in the number rules and the relationship between the coordinates.

a.

x						
y						

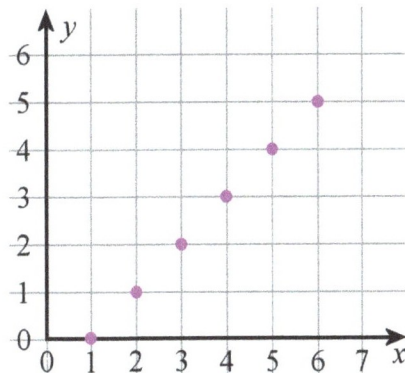

The rule for *x*-values:

Start at ____, and _____.

The rule for *y*-values:

Start at ____, and _____.

The relationship between each pair of *x* and *y:*

b. Notice that the *y*-axis is now scaled differently.

x					
y					

x				
y				

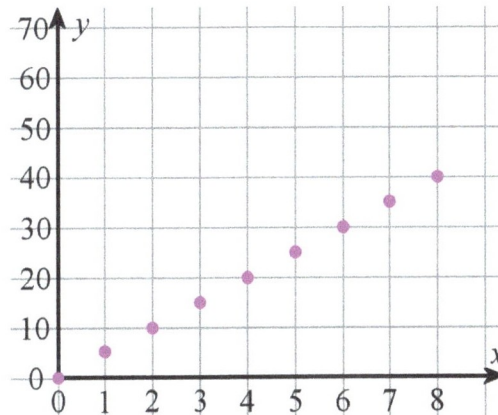

x-values: Start at ____, and _____.

y-values: Start at ____, and _____.

The relationship between each pair of *x* and *y:*

c.

x						
y						

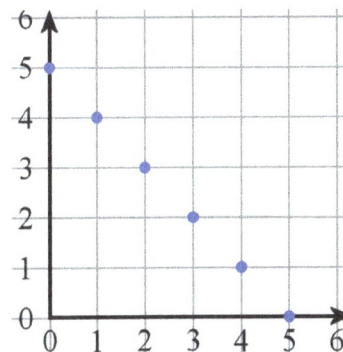

x-values: Start at ____, and _____.

y-values: Start at ____, and _____.

The relationship between each pair of *x* and *y:*

More Number Patterns in the Coordinate Grid

In each exercise, plot the points from the "number rules" in the coordinate grids.

1. _x_-values: start at 0, and add 1 each time.

   _y_-values: start at 3, and add 1 each time.

x							
y							

The rule between each _x_ and _y_-coordinate:

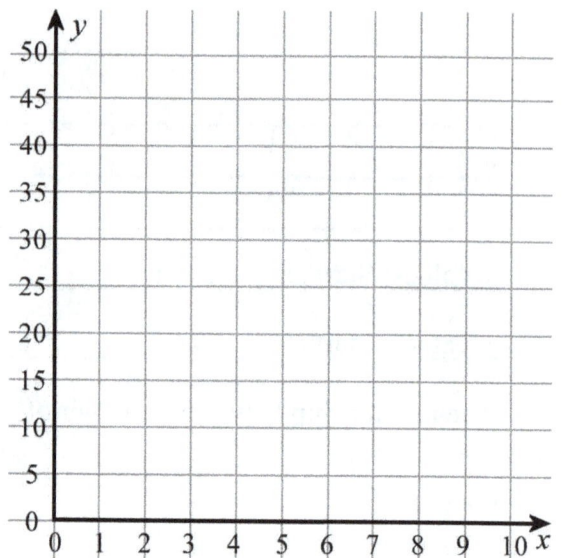

Explain in your own words why this is so.

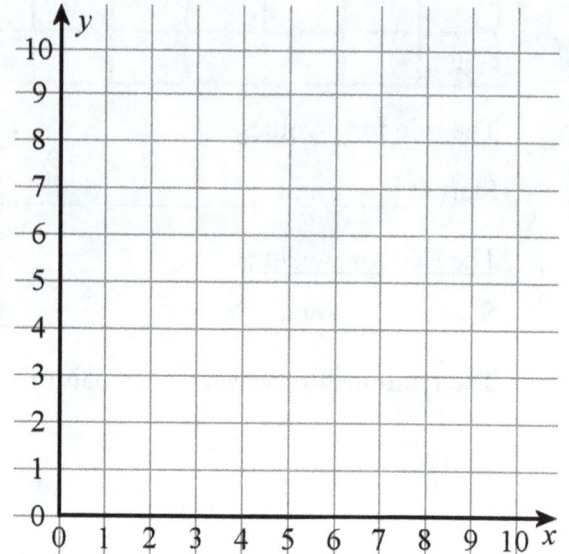

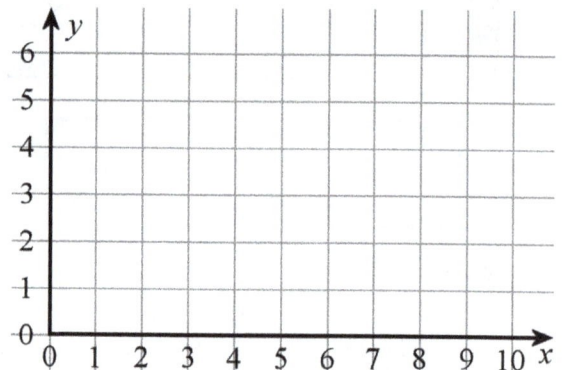

2. _x_-values: start at 0, and add 1 each time.

   _y_-values: start at 0, and add ½ each time.

x							
y							

The rule between each _x_ and _y_-coordinate:

Explain in your own words why this is so.

3. Note the scaling.

   The rule for _x_-values: start at 0, and add 1 each time.
   The rule for _y_-values: start at 0, and add 5 each time.

x							
y							

The rule between each _x_ and _y_-coordinate:

4. The rule for x-values: start at 10, and subtract 1 each time.
   The rule for y-values: start at 1, and add 2 each time.

x							
y							

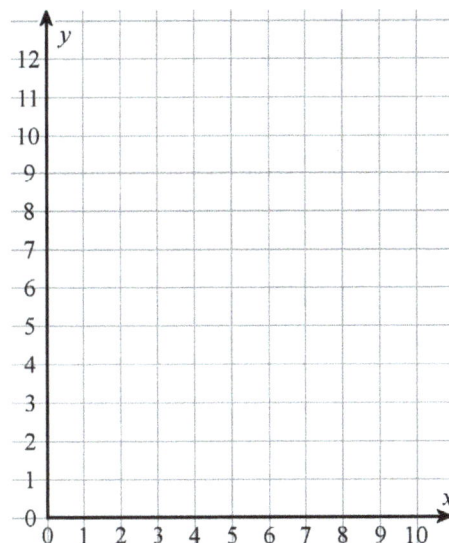

5. The rule for x-values: start at 1, and add 1 each time.
   The rule for y-values: start at 5, and subtract ½ each time.

x							
y							

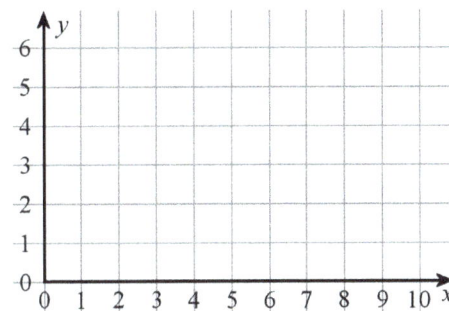

6. Make your own rule.

   The rule for x-values: start at _____, and

   The rule for y-values: start at _____, and

x							
y							

   If you can, write the rule between each x and y-coordinate:

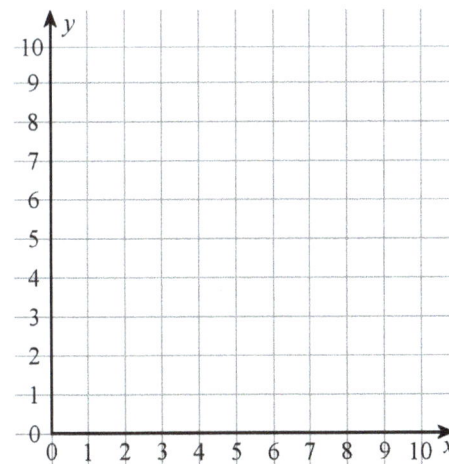

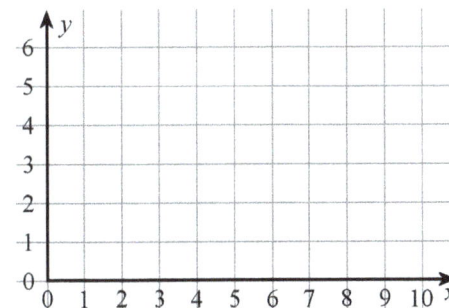

7. Make your own rule. Plot the points in the same grid as for #5 or in the small grid (if they fit).

   The rule for x-values: start at _____, and

   The rule for y-values: start at _____, and

x							
y							

   If you can, write the rule between each x and y-coordinate:

You can make and plot more number rules of your own on graph paper.

8.

x						
y						

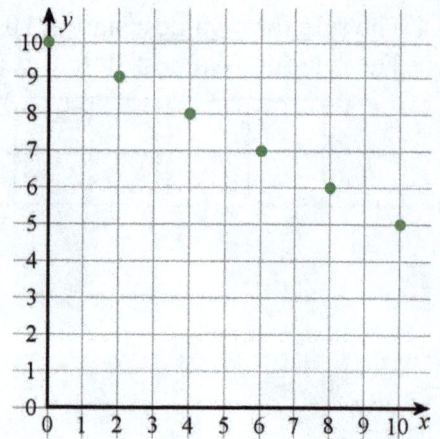

The rule for *x*-values:

Start at ____ , and _____ .

The rule for *y*-values:

Start at ____ , and _____ .

The rule between each pair of *x* and *y* may be

harder to see; it is $y = 10 - \dfrac{x}{2}$ or $\dfrac{x}{2} + y = 10$.

9. This time the coordinate grid
is *scaled* differently.

The rule for *x*-values:
start at 0, and add 10 each time.

The rule for *y*-values:
start at 2, and add 1 each time.

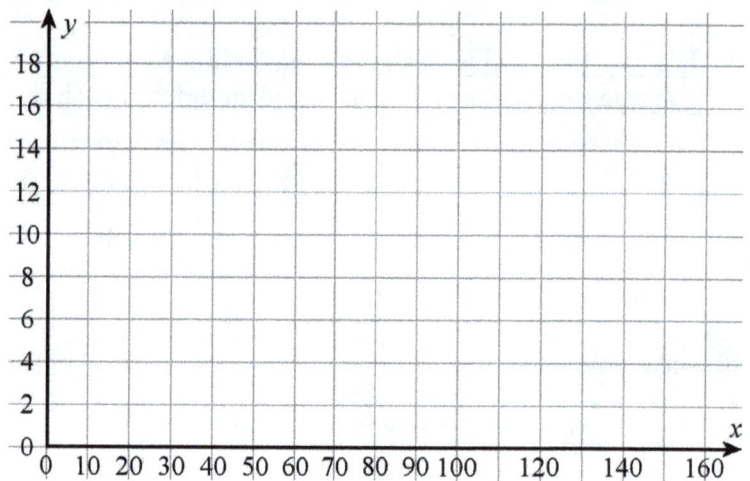

x												
y												

(Challenge) What simple rule ties the *x* and *y*-coordinates together in each case?

Puzzle Corner

x-values: start at 8, and subtract ½ each time.
y-values: start at 0, and add 1 each time.

x							
y							

Write the rule between each *x* and *y*-coordinate:

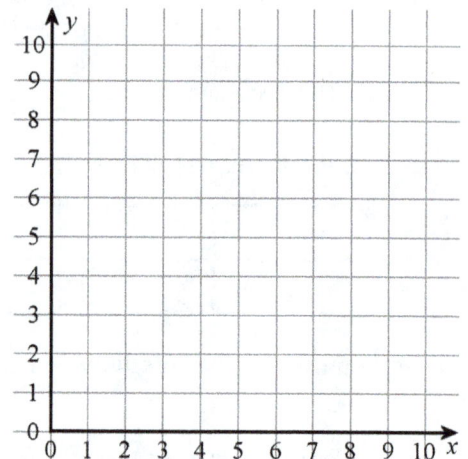

54

Review

1. Solve (without a calculator).

   a. 7,587 ÷ 27

   b. 2,829 ÷ 41

   c. 249 × 382

2. Solve 83,493 − y = 21,390.

3. Solve in the right order. You can enclose the
   operation to be done first in a "bubble" or a "cloud."

a. 5 × (3 + 8) = _____	**b.** 20 + 240 ÷ 8 + 90 = _____
c. 100 − 2 × 5 × 7 = _____	**d.** 70 − 2 × (2 + 5) = _____

4. Divide mentally, and solve in the right order.

a. $\dfrac{3636}{6}$ =	**b.** $\dfrac{3608}{4}$ =	**c.** $\dfrac{4050}{5}$ =
d. 42 + $\dfrac{255}{5}$ =	**e.** $\dfrac{4,804}{2+2}$ =	

55

5. Find a number to fit in the box so the equation is true.

a. $25 = 7 + \boxed{} \times 2$	**b.** $72 \div 8 = (6 - 3) \times \boxed{}$	**c.** $(4 + \boxed{}) \div 3 = 2 + 2$

6. Write an expression _or_ an equation to match each written sentence. You do not have to solve.

a. The difference of x and 9	**b.** The sum of y and 3 and 8 equals 28.
c. The quotient of 60 and b is equal to 12.	**d.** The product of 8, x and y

7. Which expression matches the problem? Also, solve the problem.

Three girls divided equally the cost of buying four sandwiches for $3.75 each. How much did each girl pay?

(1) $3 \times \$3.75 - 4$ **(2)** $3 \times \$3.75 \div 4$

(3) $\$3.75 \div 4 \times 3$ **(4)** $4 \times \$3.75 \div 3$

8. Write a _single_ expression (number sentence) for the problems, and solve.

a. Bonnie and Ben bought an umbrella for $12 and boots for $17, and divided the cost equally. How much did each pay?

b. Henry bought five jugs of milk for $4.50 each. In the end, the grocer gave him $2 off his bill. What did Henry pay?

9. Draw a bar model to represent the equations. Then solve them.

a. $R \div 4 = 544$

b. $4 \times R = 300$

10. Divide and indicate the remainder, if any.
    Use long division.

    a. $6{,}764 \div 81$

    b. $309{,}855 \div 46$

11. How many times can you subtract 9 from 23,391 before you "hit" zero?

12. If you spend exactly $2.25 every day to make a phone
    call, how much will those phone calls cost you in
    a year?

13. If 5,000 people need to be moved from place A to place B
    by buses, and one bus seats 46 people, how many buses
    are needed?

14. An airplane travels at a constant speed of 880 km per
    hour. *Estimate* about how long it will take for it
    to fly 5,800 km.

15. Three boxes of tea bags cost $15.90.
How much do two boxes cost?

16. Plot the points from the "number rule" on the coordinate grid. Fill in the rest of the table first, using the rule given.

The rule is: $y = 9 - x$.

x	0	1	2	3	4
y					

x	5	6	7	8	9
y					

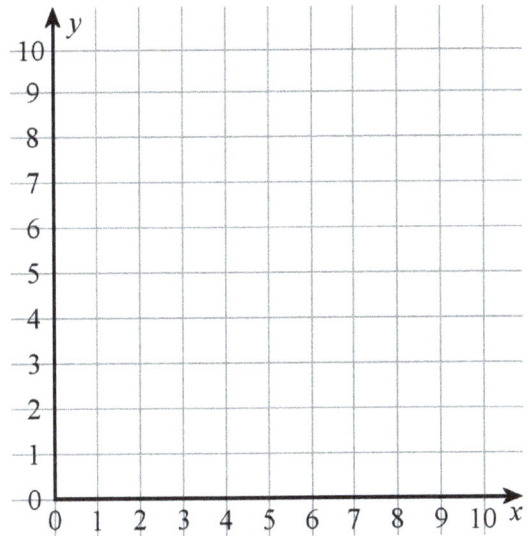

Answer Key

Warm Up: Mental Math, pp. 7-8

Page 7

1. a. 38; 75   b. 95; 85   c. 7,800; 17,720
   d. 16; 34   e. 27; 128   f. 1,000; 253

2. The total length of the track is 4 km 300 m.

3. The temperature was 88°F previously.

4. The fourth crate of apples weighs 7 kg.

5. a. 73   b. 210   c. 90

7. a.   60; 120; 180; 240; 300; 360; 420; 480; 540
   b. 1,080; 960; 840; 720; 600; 480; 360; 240; 120
   c.   130; 170; 210; 250; 290; 330; 370; 410; 450

8. $27 + $27 + $3 + $3 + $3 = $63 approximately.

9. a. 130   b. 215   c. 246
   d. 535   e. 135   f. 288
   g. 1,435   h. 633   i. 275
   j. 198   k. 128   l. 981

Page 8

6.

a. $20 \times 6 = 120$ $200 \times 6 = 1,200$ $200 \times 600 = 120,000$	b. $10 \times 35 = 350$ $100 \times 35 = 3,500$ $20 \times 100 = 2,000$	c. $400 \times 500 = 200,000$ $60 \times 80 = 4,800$ $100 \times 430 = 43,000$

The Order of Operations, pp. 9-10

Page 9

1. a. 6   b. 260   c. 31   d. 20

2. a. 3   b. 500   c. 63   d. 100

Page 10

3. a. $5 \times (30 - 9) = 105$ or $(30 - 9) \times 5 = 105$ b. $7 \times 6 + 20 = 62$ (no need for parentheses)
   c. $(14 + 15 + 16) \div 3 = 15$ d. $100 - (27 + 37) = 36$

4. a. $90 - (26 + 6) \times 2 = 26$ or $90 - 2 \times (26 + 6) = 26$
   b. $100 - 5 \times 7 + 34 = 99$
   c. $55 - 5 \times (36 \div 9) = 35$ or $55 - 36 \div 9 \times 5 = 35$

5. a. 4   b. 36   c. 45   d. 44   e. 108   f. 80

6. a. 18   b. 50   c. 3

Equations, pp. 11-12

Page 11

1. a. equation   b. expression   c. equation   d. equation   e. equation   f. expression

2. a. (3) $50 - 3 \times $8 = $26. His change was $26.   b. (2) $6 \times ($16 - $5) = $66. The total cost is $66.
   c. (4) ($8 + $13) \div 2 = $10.50. Andy's share is $10.50.   d. (1) $48 \div 4 + $30 \div 3 = $22. Melissa pays $22.

Page 12

3. a. false   b. false   c. true

   When changing one number in (a) and (b), answers vary. For example: a. $1 + \dfrac{32}{8} = 5$ b. $(6 - 2) \times 3 = 5 + 7$

4. a. $(10 + 40 + 40) \times 2 = 180$ b. $144 = 3 \times (2 + 4) \times 8$ c. $40 \times 3 = (80 - 50) \times 4$

Equations, cont.

5.

a. $40 = (11 + 9) \times 2$	b. $4 \times 8 = 5 \times 6 + 2$	c. $4 + 5 = (20 - 2) \div 2$
d. $81 = 9 \times (2 + 7)$	e. $12 \times 11 = 12 + 20 \times 6$	f. $(4 + 5) \times 3 = 54 \div 2$

6. a. $s = 330$ b. $x = 10$ c. $y = 140$

7. Answers will vary. Examples:
$3 \times 3 + 1 = 1 \times 11 - 1$
$3 \times 11 + 3 = 3 \times 3 \times 3 + 11 - 1 - 1$
$11 - 3 = 3 \times 3 - 1$

Review: Addition and Subtraction, pp. 13-15

1. a. Addition: $x + 1{,}057 = 2{,}370$
Subtraction: $2{,}370 - 1{,}057 = x$ OR $2{,}370 - x = 1{,}057$
Solution: $x = 1{,}313$

b. Addition: $3{,}938 + x + 1{,}506 = 12{,}000$
Subtraction: $12{,}000 - 3{,}938 - 1{,}506 = x$
OR $12{,}000 - x - 1{,}506 = 3{,}938$
OR $12{,}000 - x - 3{,}938 = 1{,}506$
Solution: $x = 6{,}556$

c. Addition: $2x + 200 = 560$
Subtraction: $560 - 200 = 2x$ OR $560 - 2x = 200$
Solution: $x = 180$

2. a. $68 + s$ b. $y - 37$
c. $60 + b + 40 = 120$ d. $80 - x = 35$

3. a. $20 - (7 + 5)$ b. $20 - (7 - 5)$
c. $(7 - 5) + 20$ d. $7 + (20 - 5)$

4. a. $(15 - 6) + 16$ OR $16 + (15 - 6)$
b. $100 - (5 + 80)$

5. a. $7{,}000 - (1{,}500 + 2{,}500) = 3{,}000$
$7{,}000 - 2{,}500 - 1{,}500 = 3{,}000$
$7{,}000 - (2{,}500 - 1{,}500) = 6{,}000$

The first and second had the same answer.

b. $600 + 30 - 30 + 30 - 30 = 600$
$600 - (30 + 30 + 30 + 30) = 480$
$600 - 30 - 30 - 30 - 30 = 480$

The second and third had the same answer.

6. b. $\$900 - \$14 \times \$58$

7. c. $3 \times (24 + 12)$

8. b. $9 \times \$7 \div 2$

9. a. The total cost: $15 \times \$2 + \$6 = \$36$
Change: $\$50 - (15 \times 2 + 6) = \$14$
b. $(\$9 + \$8 + \$13) \div 3 = \$10$
Each child paid \$10.
c. $(\$128 - 31) \times 5 = \$485$.
The total cost is \$485.

Review: Multiplication and Division, pp. 16-18

1. a. $4 \times 305 = w$ OR $305 \times 4 = w$
$w \div 4 = 305$ OR $w \div 305 = 4$.
Solution: $w = 1{,}220$

b. $5 \times w = 305$ OR $w \times 5 = 305$
$305 \div w = 5$ OR $305 \div 5 = w$
Solution: $w = 61$

2. a. $6 \times y = 90$; $y = 15$ b. $y \div 6 = 90$; $y = 540$

3. a.

R = 600

b.

R = 24

c.

$y = 720$

62

Review: Multiplication and Division, cont.

Page 17

4. a. 52×8 b. $15{,}000 \div 300$
   c. $4 \times S \times 18$ d. $80 \div x$
   e. $240 \div 8 = 30$ f. $3 \times 5 \times T = 60$

5. $280 \div N = 4; N = 70$

6. $H \div 91 = 3; H = 273$

Page 18

7. a. yes   b. no   c. yes
   d. yes   e. no   f. yes

8. b. matches

9. Multiply the divisor by the quotient to find the dividend.

10. Divide the dividend by the quotient to find the divisor.

11. a. $M = 4$ b. $M = 15$ c. $M = 9$
    d. $N = 2{,}200$ e. $N = 12{,}000$ f. $N = 8$

Balance Problems and Equations, Part 1, pp. 19-21

Page 19

1.

a. Equation: $9 = \boxed{} + 3$ Solution: $\boxed{} = 6$	b. Equation: $3\,\bigcirc = 21$ Solution: $\bigcirc = 7$
c. Equation: $\boxed{} + \boxed{} + 2 = 16$ Solution: $\boxed{} = 7$	d. Equation: $\boxed{} + 7 = 51$ Solution: $\boxed{} = 44$

Page 20

2.

a. $x = 24 + 7$ $x = 31$	b. $x + 12 = 38 + 5$ $x + 12 = 43$ $x = 31$

3.

a.
$x + 18 = 5 + 31$
$x + 18 = 36$
$x = 18$

b.
$8 + 17 = 11 + x$
$25 = 11 + x$
$14 = x$
$x = 14$

Balance Problems and Equations, Part 1, cont.

4.

a.	b.	c.
$88 = 2x$	$2x = 16 + 6$	$3x = 6 + 32 + 4$
$44 = x$	$2x = 22$	$3x = 42$
$x = 44$	$x = 11$	$x = 14$

5.

a.
$$3x = 16 + 35$$
$$3x = 51$$
$$x = 17$$

b.
$$2 + 27 + 25 = 6x$$
$$54 = 6x$$
$$9 = x$$
$$x = 9$$

Puzzle corner. a. $x = 463$ b. $x = 0$

Balance Problems and Equations, Part 2, pp. 22-24

1.

a. $2x + 47 = 3x$	b. $2x = x + 17$
$47 = x$	$x = 17$
$x = 47$	
c. $2x + 7 = x + 19$	d. $3x + 7 = 2x + 23$
$x + 7 = 19$	$x + 7 = 23$
$x = 12$	$x = 16$
e. $2x + 44 = 4x$	f. $5x = 2x + 24$
$44 = 2x$	$3x = 24$
$x = 22$	$x = 8$

2.

a. $3x + 9 = 27$	b. $2x + 3 = 93$
$3x = 18$	$2x = 90$
$x = 6$	$x = 45$
c. $2x + 6 = 32 + 4$	d. $36 + 7 = 5x + 13$
$2x + 6 = 36$	$43 = 5x + 13$
$2x = 30$	$30 = 5x$
$x = 15$	$6 = x$
	$x = 6$

3.

a. $x + 51 = 2x + 5$	b. $9 + x + 6 = 2x + 2$	c. $4x + 6 = x + 13 + 5$
$51 = 5 + x$	$x + 15 = 2x + 2$	$4x + 6 = x + 18$
$x = 46$	$x + 13 = 2x$	$3x + 6 = 18$
	$13 = x$	$3x = 12$
	$x = 13$	$x = 4$

Balance Problems and Equations, Part 2, cont.

Page 24

4.

a. $2x + 5 = 41$ $2x = 36$ $x = 18$	b. $3x + 37 = 4x$ $37 = x$ $x = 37$
c. $x + 15 = 2x + 7$ $15 = x + 7$ $x = 8$	d. $3x + 8 = 26$ $3x = 18$ $x = 6$

Multiplying and Dividing in Parts, pp. 25-28

Page 25

1.

a. $7 \cdot 99 = 7 \cdot (100 - 1)$ $= 700 - 7 = \underline{693}$	b. $4 \cdot 999 = 4 \cdot (\underline{1,000} - \underline{1})$ $= \underline{4,000} - 4 = \underline{3,996}$
c. $5 \cdot 104 = 5 \cdot (\underline{100} + \underline{4})$ $= \underline{500} + 20 = \underline{520}$	d. $5 \cdot 998 = 5 \cdot (\underline{1,000} - \underline{2})$ $= \underline{5,000} - 10 = \underline{4,990}$
e. $6 \cdot 98 = 6 \cdot (100 - 2)$ $= \underline{600} - 12 = \underline{588}$	f. $7 \cdot 2030 = 7 \cdot (\underline{2,000} + \underline{30})$ $= \underline{14,000} + 210 = \underline{14,210}$

2.

a. Total area: $\underline{3} \cdot (\underline{6} + \underline{4})$
The areas of the two rectangles: $\underline{3} \cdot \underline{6}$ and $\underline{3} \cdot \underline{4}$

b. Total area: $\underline{4} \cdot (\underline{5} + \underline{4})$
The areas of the two rectangles: $\underline{4} \cdot \underline{5}$ and $\underline{4} \cdot \underline{4}$

c. Total area: $\underline{6} \cdot (\underline{6} + \underline{8})$
The areas of the two rectangles: $\underline{6} \cdot \underline{6}$ and $\underline{6} \cdot \underline{8}$

d. Total area: $\underline{5} \cdot (\underline{2} + \underline{3})$
The areas of the two rectangles: $\underline{5} \cdot \underline{2}$ and $\underline{5} \cdot \underline{3}$

Page 26

3. a. 80 is the partial product of $10 \cdot 8$
   (10 from 16 and 8 from 78).

   700 is the partial product of $10 \cdot 70$
   (10 from 16 and 70 from 78).

b.

		5	6
X		8	4
		2	4
	2	0	0
	4	8	0
4	0	0	0
4	7	0	4

c.

		1	7
X		9	5
		3	5
		5	0
	6	3	0
	9	0	0
1	6	1	5

65

Multiplying and Dividing in Parts, cont.

4.

a. $29 \cdot 17$

200	90
140	63

$29 \cdot 17 = 20 \cdot 10 + 20 \cdot 7$
$+ 9 \cdot 10 + 9 \cdot 7$
$= 200 + 140 + 90 + 63 = 493$

b. $75 \cdot 36$

2100	150
420	30

$75 \cdot 36 = 70 \cdot 30 + 70 \cdot 6$
$+ 5 \cdot 30 + 5 \cdot 6$
$= 2{,}100 + 420 + 150 + 30 = 2{,}700$

Page 27

5.

a. $\dfrac{80}{2} + \dfrac{12}{2} = 40 + 6 = 46$	b. $\dfrac{350}{5} + \dfrac{15}{5} = 70 + 3 = 73$	c. $\dfrac{400}{4} - \dfrac{12}{4} = 100 - 3 = 97$
d. $\dfrac{9{,}300}{3} - \dfrac{60}{3} = 3{,}100 - 20 = 3{,}080$	e. $\dfrac{350}{7} + \dfrac{21}{7} - \dfrac{7}{7} = 50 + 3 - 1 = 52$	f. $\dfrac{900}{9} - \dfrac{18}{9} = 100 - 2 = 98$
g. $\dfrac{22 \text{ ft}}{2} + \dfrac{9 \text{ in.}}{2} = 11 \text{ ft } 4.5 \text{ in.}$	h. $\dfrac{40 \text{ kg}}{5} + \dfrac{750 \text{ g}}{5} = 8 \text{ kg} + 150 \text{ g}$	i. $\dfrac{12 \text{ L}}{4} + \dfrac{600 \text{ ml}}{4} = 3 \text{ L } 150 \text{ ml}$

6. a. 206   b. 203   c. 103   d. 201   e. 502

7. One way: Two liters and 250 milliliters equal 2,000 ml + 250 ml = 2,250 ml.
Then, 2,250 ml ÷ 4 = 562 ½ ml ≈ <u>560 ml per person.</u>

Another way: Two liters divided among 4 people is half a liter (500 ml), each. Then, 250 milliliters divided among 4 people is 250/4 = 62 ½ ml per person. So each of the four people at the party gets 500 ml plus 62 ½ ml, or 500 + 62 ½ = 562 ½ ml ≈ <u>560 ml per person.</u>

Page 28

8.

a. $\dfrac{15}{5} + \dfrac{4}{5} = 3\dfrac{4}{5}$	b. $\dfrac{44}{11} + \dfrac{7}{11} = 4\dfrac{7}{11}$
c. $\dfrac{6}{7} + \dfrac{70}{7} = 10\dfrac{6}{7}$	d. $\dfrac{420}{6} + \dfrac{2}{6} = 70\dfrac{2}{6}$
e. $\dfrac{240}{4} + \dfrac{12}{4} + \dfrac{3}{4} = 60 + 3 + \dfrac{3}{4} = 63\dfrac{3}{4}$	f. $\dfrac{2}{9} + \dfrac{36}{9} + \dfrac{270}{9} = 4 + 30 + \dfrac{2}{9} = 34\dfrac{2}{9}$

9. a. 100 3/4   b. 303 2/3   c. 1,004 4/5   d. 20 1/4   e. 42 1/3   f. 60 5/6

10. a. 7 ÷ 14 = 1/2   b. 7 ÷ 21 = 1/3   c. 80 ÷ 11 = 7 3/11
    d. 6/8 + 3 + 30 = 33 6/8   e. 117 ÷ 4 = 29 1/4   f. 100 ÷ 30 = 3 1/3

Puzzle corner:

a. $\dfrac{250 - 3}{10} = 25 - \dfrac{3}{10}$

b. $\dfrac{11 - 3}{5} = 2\dfrac{1}{5} - \dfrac{3}{5}$

Page 29

1.

a. $410 + 2 \times 19$	b. $3 \times 50 + 4 \times 150$	c. $70 \times 80 - 40 \times 50$
$= 410 + 38 = \underline{448}$	$= 150 + 600 = \underline{750}$	$= 5,600 - 2,000 = \underline{3,600}$
d. $14 + (530 - 440)$	e. $45 + 56 + 35$	f. $300 \div 5 - 400 \div 10$
$= 14 + 90 = \underline{104}$	$= 101 + 35 = \underline{136}$	$= 60 - 40 = \underline{20}$

2. a. 93   b. 655   c. 380

3. a. 60   c. 8   e. 560   g. 21   i. 6
   b. 72   d. 50   f. 40   h. 200   j. 9

Page 30

4.

a. $500 - 40 - 3 \times 50$ $= 460 - 150$ $= \underline{310}$	b. $1,020 - (40 - 10) \times 20$ $= 1,020 - 30 \times 20$ $= 1,020 - 600 = \underline{420}$
c. $42,000 - 12,000 + 3 \times 5,000$ $= 30,000 + 15,000 = \underline{45,000}$	d. $(70 - 20) \times 70$ $= 50 \times 70 = \underline{3,500}$
e. $\dfrac{210}{2} + 3 \times 15$ $= 105 + 45 = \underline{150}$	f. $250 \times 4 + \dfrac{6,300}{70}$ $= 1,000 + 90 = \underline{1,090}$

5. a. $x = 2,800$ b. $M = 60$ c. $y = 180$

6.

n	130	250	360	410	775	820	1,000
$n - \underline{35}$	95	215	325	375	740	785	965

7.

n	3	5	12	15	25	35	60
$n \times \underline{40}$	120	200	480	600	1,000	1,400	2,400

8. a. <u>Each piece of board is 110 cm long</u>: $(600 \text{ cm} - 50 \text{ cm}) \div 5 = 550 \text{ cm} \div 5 = 110 \text{ cm}$.

9. a. Evelyn's hourly wage is $\$104.00 \div 8 = \underline{\$13.00 \text{ per hour}}$.
   b. Evelyn earns $\$104 \times 5 = \underline{\$520 \text{ in a week}}$, and $\$520 \times 13 = \underline{\$6,760 \text{ in three months}}$.

10. a.

6 3/4 cups of flour	4 1/2 teaspoons of cinnamon
9 teaspoons of baking powder	2 1/4 teaspoons of nutmeg
1 cup of honey	1 1/2 teaspoons of ground cloves
1 1/2 cups of butter	2 1/4 cups of walnuts

   b. They baked 7 ½ dozen biscuits.

Page 31

1. a.
```
        2 4
        5 3 6
    x       7 1
        5 3 6
    3 7 5 2
    3 8 0 5 6
```
b.
```
        3 4 6
    $ 2 4.5 9
    x       7 0
    $ 1 7 2 1.3 0
```
c.
```
        2 0 6
    x   9 1 5
    1 0 3 0
        2 0 6
    1 8 5 4
    1 8 8 4 9 0
```

d.
```
       1 3 8 7
   4)5 5 4 8
     4
     1 5
     1 2
       3 4
       3 2
         2 8
         2 8
          0
```
e.
```
       8 5.6
   7)5 9 9.2
     5 6
       3 9
       3 5
         4 2
         4 2
          0
```
f.
```
       1 0 3 4
   8)8 2 7 2
     8
     0 2 7
       2 4
         3 2
         3 2
          0
```

2. You multiply the quotient by the divisor to check your division.

```
    1 3 2
    1 3 8 7
x         4
    5 5 4 8
```
```
      3 4
      8 5.6
x         7
    5 9 9.2
```
```
      2 3
    1 0 3 4
x         8
    8 2 7 2
```

Page 32

3. a. 6,048   b. 34.95   c. 109,841 R2

4.

a. 437 ÷ 6 = 72 R5	b. 2,045 ÷ 3 = 681 R1
6 × 72 + 5 = 432 + 5 = 437 (It checks.)	3 × 681 + 1 = 2,043 + 1 = 2,044. It does not check. The correct answer is 2,045 ÷ 3 = 681 R2.

Page 33

5. She noticed that the remainder, 9, was more than the divisor, 8. In reality they can get one more bag of buns, and have only one bun left over.

6. We can first of all divide and get 542 ÷ 25 = 21 R17. This means 21 classes of 25 students, and 17 students left over. These 17 students need to be spread into 17 classes, one per class. So, in the end we get 17 classes with 26 students, and four classes with 25 students.

Page 33

7.

$2 \times 45 = 90$ $3 \times 45 = 135$ $4 \times 45 = 180$ $5 \times 45 = 225$ $6 \times 45 = 270$ $7 \times 45 = 315$ $8 \times 45 = 360$ $9 \times 45 = 405$	$\begin{array}{r} 8\ 9 \\ \text{a. } 45\overline{)4\ 0\ 0\ 5} \\ \underline{-3\ 6\ 0} \\ 4\ 0\ 5 \\ \underline{-4\ 0\ 5} \\ 0 \end{array}$	$\begin{array}{r} 3\ \ 4 \\ 8\ 9 \\ \times\ \ \ 4\ 5 \\ \hline 4\ 4\ 5 \\ +3\ 5\ 6\ 0 \\ \hline 4\ 0\ 0\ 5 \end{array}$
$2 \times 75 = 150$ $3 \times 75 = 225$ $4 \times 75 = 300$ $5 \times 75 = 375$	$\begin{array}{r} 0.2\ 6\ 5 \\ \text{b. } 75\overline{)1\ 9.8\ 7\ 5} \\ \underline{-1\ 5\ 0} \\ 4\ 8\ 7 \\ \underline{-4\ 5\ 0} \\ 3\ 7\ 5 \\ \underline{-3\ 7\ 5} \\ 0 \end{array}$	$\begin{array}{r} 0.2\ 6\ 5 \\ \times\ \ \ \ 7\ 5 \\ \hline 1\ 3\ 2\ 5 \\ +1\ 8\ 5\ 5\ 0 \\ \hline 1\ 9.8\ 7\ 5 \end{array}$

Page 34

8. a. 1,813 R1; $48 \times 1{,}813 + 1 = 87{,}025$
   b. 9,685 R10; $90 \times 9{,}685 + 10 = 871{,}660$
   c. 658 R66; $82 \times 658 + 66 = 54{,}022$

Page 35

9.

a. 2,960 R86 $\begin{array}{r} 2\ 9\ 6\ 0 \\ 101\overline{)2\ 9\ 9\ 0\ 4\ 6} \\ \underline{-2\ 0\ 2} \\ 9\ 7\ 0 \\ \underline{-9\ 0\ 9} \\ 6\ 1\ 4 \\ \underline{-6\ 0\ 6} \\ 8\ 6 \end{array}$	$\begin{array}{r} 3,4 \\ 2\ 9\ 6\ 0 \\ \times\ 1\ 0\ 1 \\ \hline 2\ 9\ 6\ 0 \\ 0 \\ +2\ 9\ 6\ 0\ 0\ 0 \\ \hline 2\ 9\ 8\ 9\ 6\ 0 \\ +\ \ \ \ \ \ \ 8\ 6 \\ \hline 2\ 9\ 9\ 0\ 4\ 6 \end{array}$
b. 29,546 R48 $\begin{array}{r} 2\ 9\ 5\ 4\ 6 \\ 123\overline{)3\ 6\ 3\ 4\ 2\ 0\ 6} \\ \underline{-2\ 4\ 6} \\ 1\ 1\ 7\ 4 \\ \underline{1\ 1\ 0\ 7} \\ 6\ 7\ 2 \\ \underline{-6\ 1\ 5} \\ 5\ 7\ 0 \\ \underline{-4\ 9\ 2} \\ 7\ 8\ 6 \\ \underline{-7\ 3\ 8} \\ 4\ 8 \end{array}$	$\begin{array}{r} 2\ 9\ 5\ 4\ 6 \\ \times\ \ 1\ 2\ 3 \\ \hline 8\ 8\ 6\ 3\ 8 \\ 5\ 9\ 0\ 9\ 2\ 0 \\ +2\ 9\ 5\ 4\ 6\ 0\ 0 \\ \hline 3\ 6\ 3\ 4\ 1\ 5\ 8 \\ +\ \ \ \ \ \ \ \ 4\ 8 \\ \hline 3\ 6\ 3\ 4\ 2\ 0\ 6 \end{array}$

Review of the Four Operations 1, cont.

9. (continued)

c. 21,862 R300

```
           2 1 8 6 2
    350 ) 7 6 5 2 0 0 0
         -7 0 0
           6 5 2
          - 3 5 0
           3 0 2 0
          -2 8 0 0
             2 2 0 0
            -2 1 0 0
               1 0 0 0
              - 7 0 0
                 3 0 0
```

```
          2 1 8 6 2
       ×      3 5 0
      _____
                  0
        1 0 9 3 1 0 0
      + 6 5 5 8 6 0 0
      _____
          7 6 5 1 7 0 0
    +             3 0 0
      _____
          7 6 5 2 0 0 0
```

10.
I $42,408 \div 76 = 558$ **E** $44,217 \div 51 = 867$ **E** $128,316 \div 111 = 1,156$
M $85,104 \div 54 = 1,576$ **I** $223,496 \div 91 = 2,456$ **E** $51,313 \div 97 = 529$
O $23,530 \div 26 = 905$ **I** $30,624 \div 33 = 928$ **M** $880,341 \div 309 = 2,849$
R $61,880 \div 35 = 1,768$ **R** $133,140 \div 70 = 1,902$ **T** $113,168 \div 88 = 1,286$
V $51,944 \div 86 = 604$ **S** $11,880 \div 22 = 540$ **R** $693,360 \div 810 = 856$

What is as round as a dishpan, and no matter the size, all the water in the ocean cannot fill it up? <u>SIEVE</u>

What flies without wings? <u>TIME</u>

I am the only thing that always tells the truth. I show off everything that I see. <u>MIRROR</u>

G $200,196 \div 201 = 996$ **R** $617,105 \div 415 = 1,487$ **O** $1,388,740 \div 230 = 6,038$
O $324,729 \div 57 = 5,697$ **S** $2,863,250 \div 250 = 11,453$ **P** $759,290 \div 70 = 10,847$
E $339,388 \div 31 = 10,948$ **T** $1,049,664 \div 88 = 11,928$ **I** $678,040 \div 506 = 1,340$
S $2,337,820 \div 205 = 11,404$ **H** $236,215 \div 35 = 6,749$ **T** $250,536 \div 44 = 5,694$
E $28,548 \div 18 = 1,586$ **F** $97,920 \div 16 = 6,120$ **F** $239,397 \div 199 = 1,203$

From what heavy seven-letter word can you take away two letters and have eight left? <u>FREIGHT</u>

The more of them you take, the more you leave behind. What are they? <u>FOOTSTEPS</u>

Review of the Four Operations 2, pp. 37-39

1. a. $\$29,600 + \$13,500 + \$8,300 = \$51,400$. $\$51,400 \div 4 = \$12,850$.
   <u>The family used $12,850 for groceries.</u>

   b. $1/5 + 1/4 = 4/20 + 5/20 = 9/20$. <u>The family had 11/20 of their income left after taxes and groceries.</u>

2. a. $100 - 29.5 \times 2.6 = 100 - 76.7 = 23.3$ b. $2.3 + 9.356 + 0.403 + 908.8 = 920.859$
   c. $800 - (12.48 - 2.9) = 800 - 9.58 = 790.42$ d. $559.50 \div 3 = 186.5$

3. $4,958 \div 13 = 381$ R5   OR $4,958 \div 381 = 13$ R5

4. a. You would need to add *four* zeroes so that you can calculate the dividend to four decimal digits. You will need four decimal digits in order to round it to three decimal digits.
   b. $65.0000 \div 7 = 9.2857$, which rounds to <u>9.286.</u>

Review of the Four Operations 2, cont.

Page 38

5. 10 m × 12 m = 120 m^2 ; 120 m^2 ÷ 9 = 13.33 m^2. <u>The area of each section is 13.33 m^2.</u>

6. <u>The farmer needed 262 boxes to pack the apples.</u> Notice the problem doesn't give you *how many apples* there were, but instead tells you how many <u>kilograms</u> of apples there were. Since four apples make a kilogram, he had 2,350 × 4 = 9,400 apples. Now divide: 9,400 ÷ 36 = 261 R4. He needed 262 boxes.

Page 39

7. a.

Miles	9	18	27	54	108	135	162
Time	10 min	20 min	30 min	1 hour	2 hours	2 ½ hours	3 hours

   b. They will travel 486 miles.

   c. It will take them approximately ten hours to travel 550 miles.

8. a. At 40 mph, it takes him 1.5 minutes to drive each mile. You can solve this in many ways. For example, since he drives 40 miles in 60 minutes, you can make a table like in exercise 7, and find that he drives 20 miles in 30 minutes, 10 miles in 15 minutes, and 5 miles in 7 1/2 minutes.
   b. Twenty times as long as what it takes him to drive 5 miles: 20 × 7.5 minutes = 150 minutes.
   c. To drive 30 miles will take him six times as long as to drive 5 miles, so it takes him 6 × 7.5 minutes = 45 minutes. Dad would have to leave at 8:15 a.m. to arrive at 9 a.m.

9. a. One gallon is 128 liquid ounces. Ninety-six gallons is 96 × 128 oz = 12,288 oz. and 12,288 ÷ 8 = 1,536. They filled 1,536 eight-ounce bottles with fruit juice.
   b. $3,072 ÷ 1,536 = $2. They would have to charge at least $2 per bottle to break even.

Puzzle corner: a. 4,392 − <u>293</u> + 293 = 4,392   b. 384 ÷ 8 × <u>8</u> = 384   c. $\dfrac{1,568}{49} \times \underline{49} = 1,568$

Lessons in Problem Solving, pp. 40-43

Page 40

1. One small carpet costs $55.50 ÷ 5 × 2 = $22.20. Two of them cost $44.40.
$50 − $44.40 = $5.60 <u>His change was $5.60.</u>

2. The smaller ones hold 0.75 L ÷ 10 × 7 = 0.525 L.
Four large containers would hold 4 × 0.75 = 3 liters.
Five small containers would hold 5 × 0.525 = 2.625 liters.
In total, they hold 3 L + 2.625 L = 5.625 L. <u>So, yes, five liters of soup will fit into four large and five small containers.</u>

Page 41

3. Converting the 25 kg and 15 kg into grams: 25,000 g ÷ 20 = 1,250 g or 1.25 kg. One bag of bolts weighs 1.25 kg.
15,000 g ÷ 20 = 750 g or 0.75 kg. One bag of nuts weighs 0.75 kg.
1.25 kg + 0.75 kg = 2 kg. <u>Together the nuts and bolts weigh 2 kg.</u>

4. a. 670 ÷ 4 × 3 = 502.5 g. <u>A medium jar holds about 503 grams.</u>
   502.5 ÷ 3 × 2 = 335 g. <u>A small jar holds 335 grams.</u>
   b. The total weight is 670 g + 503 g + 335 g = 1,508 g or 1.508 kg.

Page 42

5. John initially had $30.60 ÷ 5 × 9 = $55.08.
Karen initially had $30.60 ÷ 3 × 7 = $71.40.
$71.40 − $55.08 = $16.32. <u>Karen had $16.32 more than John initially.</u>

Lessons in Problem Solving, cont.

Page 43

6. a. 569 ÷ 43 = 13 R10. So, they need 14 buses. (Of which 13 will be full, and the 14th bus will have 10 people in it.)
   b. The total mileage is 60 miles × 14 buses = 840 miles.
      The total cost is 840 × $2.15 = $1,806.

7. The original price of the first washer is $360 ÷ 9 × 10 = $400.

   The original price of the second washer is $348 ÷ 3 × 5 = $580.

   $580 − $400 = $180   <u>There was a
   difference of $180 between the
   original prices of the two washers.</u>

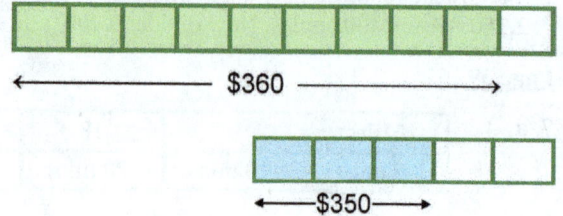

Puzzle corner:

a. Divide 1 by 10, and it gives you 0.1
   Divide 81 by 100, and it gives you 0.81.
   Divide 492 by 1,000, and it gives you 0.492.
   Divide 355 by 100, and it gives you 3.55.
   If the number has tenths, divide it by 10, if it has hundredths, divide it by 100, and if it has thousandths, divide it
   by 1,000, etc.
b. 138 ÷ 100 × 039 ÷ 100 = 0.5382.

The Coordinate Grid, pp. 44-46

Page 44

1. A (1, 2)   B (3, 4)   C (2, 9)   D (6, 5)
   E (8, 3)   F (8, 8)   G (10, 9)   H (10, 1)

Page 45

2.
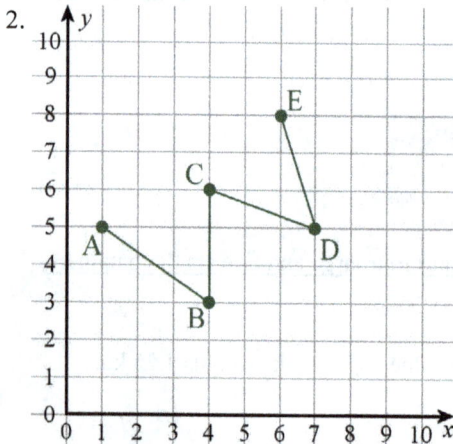

3.
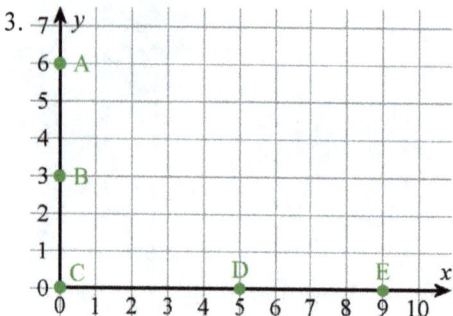

Page 45

4. a. E(4, 2) F(0, 2) G(0, 4)

   b.
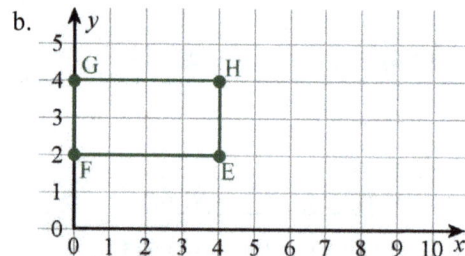

   c. H(4, 4)

Page 46

5. a. A (3, 9); B (8 1/2, 2); C (9 1/2, 5)

   b.
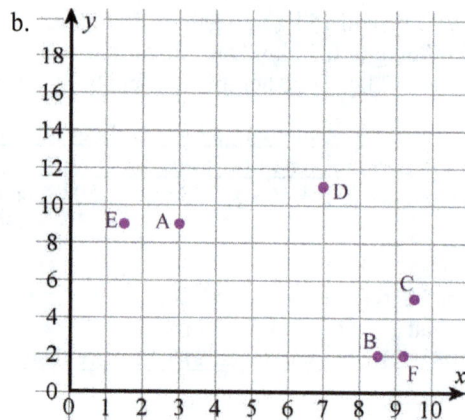

The Coordinate Grid, cont.

6. a. Answers will vary. Check how the student(s) did the scaling.

   b. Using a scaling that goes by 4s, we get:

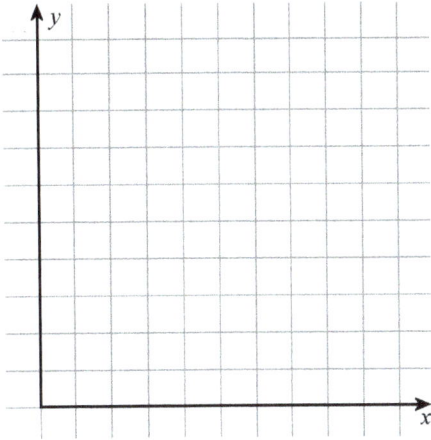

   The shape is a kite.

7. A house:

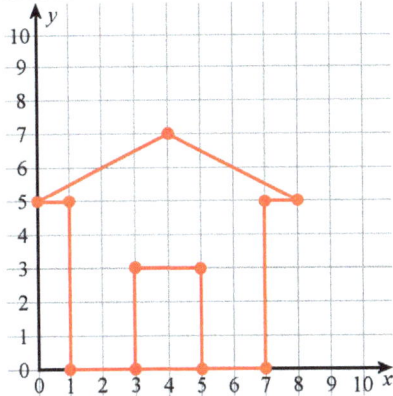

The Coordinate Grid, Part 2, pp. 47-48

1.

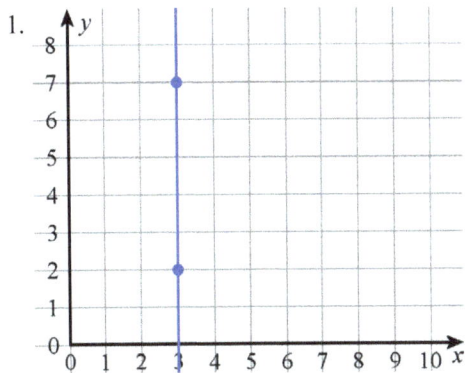

Points (3, 75) and (3, 37) would be on this line also, if it was extended.

2.

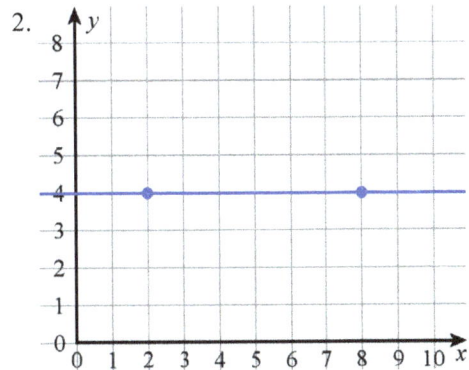

The point (35, 4) would also be on the line.

3. For all these points, the x-coordinate is 2.

4.

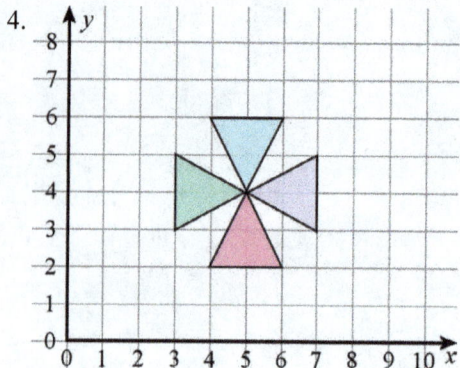

5. a. Answers will vary; check the student's work. For example, gridlines that go by tens will work, but other ways are possible, too.

b. The fourth vertex is (40, 50).

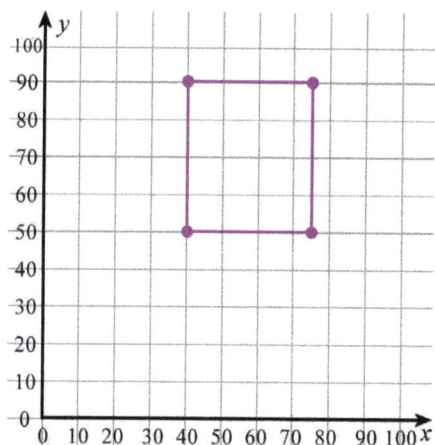

c. The width of the one side is 35 units, and of the other, 40 units. The area is $35 \times 40 = 1400$ square units.

6.

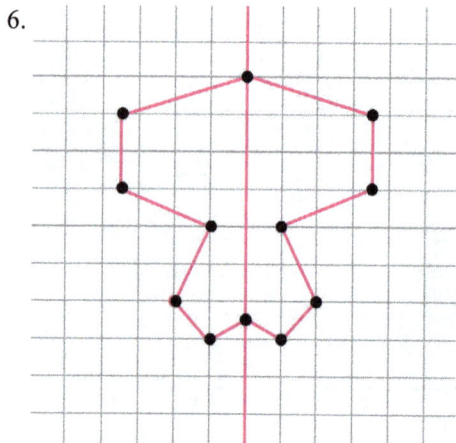

7. Answers will vary; check the student's design.

Number Patterns in the Coordinate Grid, pp. 49-51

Page 49

1. a.

x	0	1	2	3	4	5
y	0	2	4	6	8	10

b. See the image on the right.
c. $y = 2x$ or $x = y/2$
d. Answers will vary; check the student's answer. For example:
Since the x-values count by ones, and the y-values skip-count by 2s, and both start from zero, each y-value ends up being twice the corresponding x-value.

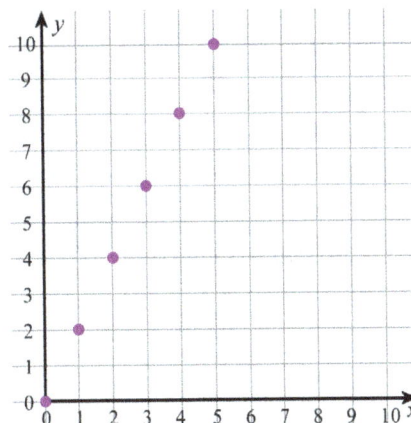

Page 50

2. a.

x	0	2	4	6	8	10
y	0	1	2	3	4	5

b. $y = x/2$ or $2y = x$

c. Answers will vary; check the student's answer. For example:
Since the y-values count by ones, and the x-values skip-count by 2s, and both start from zero, each x-value ends up being twice the corresponding y-value.

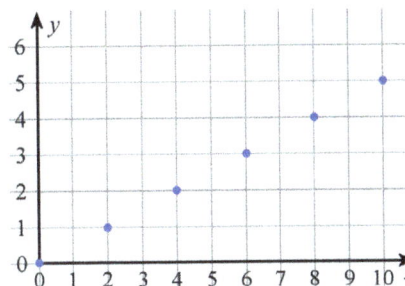

3. a.

x	0	1	2	3	4	5	6
y	6	5	4	3	2	1	0

b. $x + y = 6$ or $y = 6 - x$

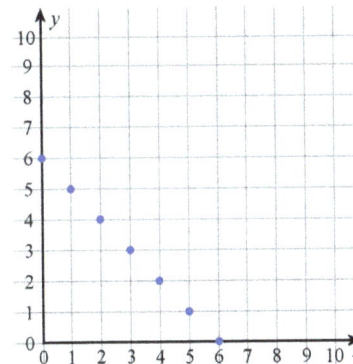

Page 51

4. a.

x	1	2	3	4	5	6
y	0	1	2	3	4	5

x-values: start at 1, and add 1 each time.
y-values: start at 0, and add 1 each time
Relationship: $y = x - 1$ or $x = y + 1$

b.

x	0	1	2	3	4	5	6	7	8
y	0	5	10	15	20	25	30	35	40

x-values: start at 0, and add 1 each time.
y-values: start at 0, and add 5 each time
Relationship: $y = 5x$ or $x = y/5$

c.

x	0	1	2	3	4	5
y	5	4	3	2	1	0

x-values: start at 0, and add 1 each time.
y-values: start at 5, and subtract 1 each time
Relationship: $y + x = 5$ or $y = 5 - x$

75

More Number Patterns in the Coordinate Grid, pp. 52-54

Page 52

1.

x	0	1	2	3	4	5	6
y	3	4	5	6	7	8	9

Relationship: $y = x + 3$ or $x = y - 3$

Explanations will vary. For example:
Both x and y values are counting by ones, but y-values started at 3,
so they are always 3 units ahead (or 3 more) of the x-values.

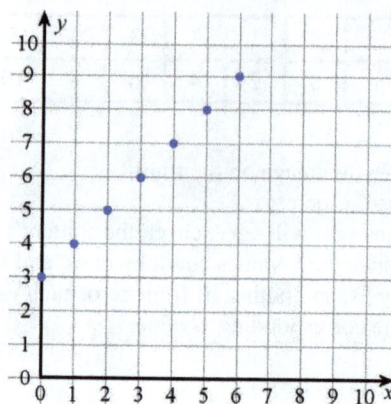

2.

x	0	1	2	3	4	5	6
y	0	½	1	1½	2	2½	3

Relationship: $y = x/2$ (or $x = 2y$)

Explanations will vary. For example:
The x-values count by ones, whereas the y-values count by halves, which
means each y-value ends up being half of the corresponding the x-value.

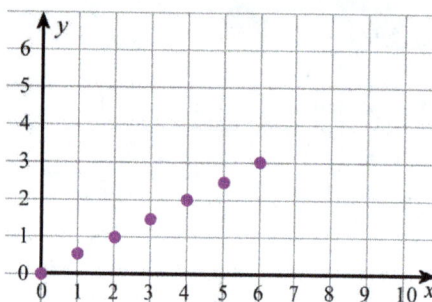

3.

x	0	1	2	3	4	5	6
y	0	5	10	15	20	25	30

Rule: $y = 5x$ (or $x = y/5$)

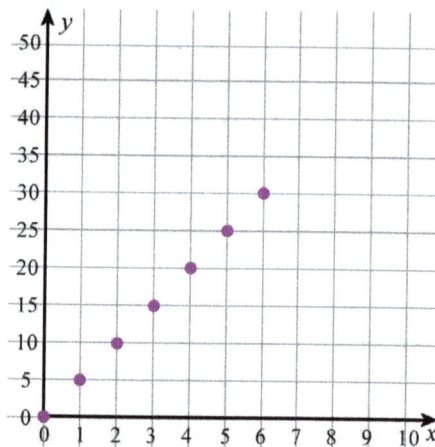

Page 53

4.

x	10	9	8	7	6	5	4
y	1	3	5	7	9	11	13

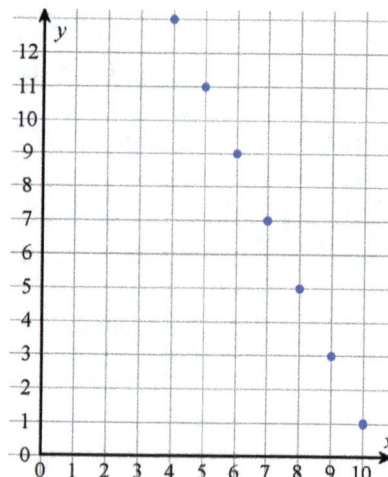

More Number Patterns in the Coordinate Grid, cont.

Page 53

5.

x	1	2	3	4	5	6	7
y	5	4 ½	4	3 ½	3	2 ½	2

The question does not ask for the relationship between x and y values, but it is this: $y = -x/2 + 5.5$ or, written differently: $y = 5.5 - 0.5x$.

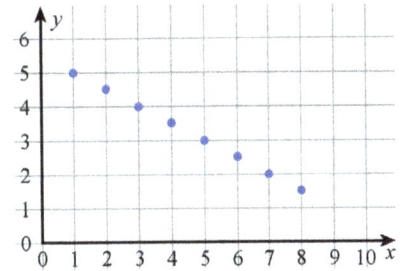

6. and 7. Answers will vary. Check the student's answer.

Page 54

8.

x	0	2	4	6	8	10
y	10	9	8	7	6	5

The rule for x-values: Start at 0 and add 2 each time.
The rule for y-values: Start at 10, and subtract 1 each time.

9.

x	0	10	20	30	40	50	60	70	80	90	100	110
y	2	3	4	5	6	7	8	9	10	11	12	13

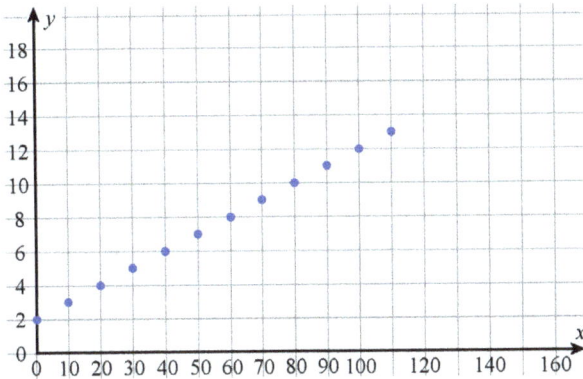

Rule: $y = x/10 + 2$

Puzzle corner.
The rule for x-values: start at 8, and subtract ½ each time.
The rule for y-values: start at 0, and add 1 each time.

x	8	7 ½	7	6 ½	6	5 ½	5	4 ½
y	0	1	2	3	4	5	6	7

The relationship: $y = -2x + 16$

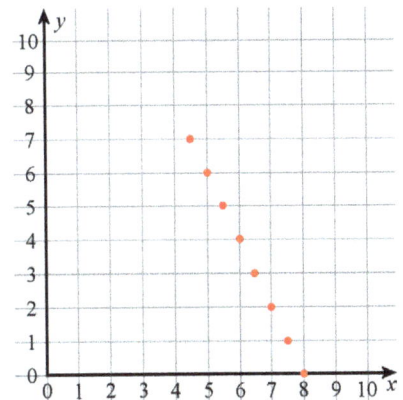

Page 55

1. a. 281   b. 69   c. 95,118

2. 83,493 − 21,390 = 62,103

3. a. 55   b. 140   c. 30   d. 56

4. a. 606   b. 902   c. 810   d. 93   e. 1,201

Page 56

5. a. 9   b. 3   c. 8

6. a. $x − 9$ b. $y + 3 + 8 = 28$ c. $60 ÷ b = 12$ d. $8 × x × y$

7. **(4)** 4 × $3.75 ÷ 3 = $5. <u>Each girl paid $5.</u>

8. a. (12 + 17) ÷ 2 = $14.50. <u>Each paid $14.50.</u>
   b. 5 × 4.50 − 2 = $20.50. <u>Henry paid $20.50.</u>

Page 57

9. a. R ÷ 4 = 544; R = 2,176   b. 4 × R = 300; R = 75

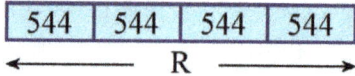   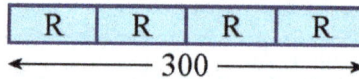

10. a. 83 R41   b. 6,735 R45

11. 23,391 ÷ 9 = 2,599 times

Page 58

12. You will spend 365 × $2.25 = <u>$821.25 in a year on phone calls.</u>

13. 5,000 ÷ 46 = 108 R32. <u>They will need 109 buses.</u>

14. Multiply to estimate, and use 900 km, instead of 880 km. Since 6 × 900 km = 5,400 and 7 × 900 km = 6,300 km, <u>it will take about 6 1/2 hours to travel 5,800 km.</u>

Page 59

15. $15.90 ÷ 3 × 2 = $10.60. <u>Two boxes of tea bags cost $10.60.</u>

16. The rule is: $y = 9 − x$.

x	0	1	2	3	4
y	9	8	7	6	5

x	5	6	7	8	9
y	4	3	2	1	0

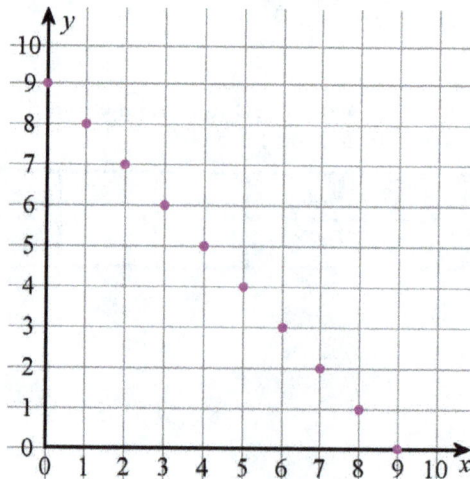

More from math MAMMOTH

Math Mammoth has a variety of resources to fit your needs. All are available as economical downloads, and most also as printed copies.

- **Math Mammoth Light Blue Series**
  A complete curriculum for grades 1-7. Each grade level includes two student worktexts (A and B), which contain all the instruction and exercises all in the same book, answer keys, tests, cumulative reviews, and a worksheet maker. International (all metric), Canadian, and South African versions are also available.

  https://www.MathMammoth.com/complete-curriculum

  https://www.MathMammoth.com/international/international

  https://www.MathMammoth.com/canada/

  https://www.MathMammoth.com/south_africa/

- **Math Mammoth Skills Review Workbooks**
  These workbooks are intended to be used alongside the Light Blue series full curriculum, and they provide additional review to the topics studied in the main curriculum, in a spiral manner.
  https://www.MathMammoth.com/skills_review_workbooks/

- **Math Mammoth Blue Series**
  Blue Series books are topical worktexts for grades 1-7, containing both instruction and exercises. The topics cover all elementary mathematics from 1st through 7th grade. These books are not tied to grade levels, and are thus great for filling in gaps.
  https://www.MathMammoth.com/blue-series

- **Make It Real Learning**
  These activity workbooks concentrate on answering the question, "Where is math used in real life?" The series includes various workbooks for grades 3-12.
  https://www.MathMammoth.com/worksheets/mirl/

- **Review Workbooks**
  Workbooks for grades 1-7 that provide a comprehensive review of one grade level of math—for example, for review during school break or summer vacation.
  https://www.MathMammoth.com/review_workbooks/

Free gift!

- Receive over 350 free sample pages and worksheets from my books, plus other freebies:
  https://www.MathMammoth.com/worksheets/free

Lastly...

- Inspire4 is an inspirational website for the whole family I've been privileged to help with:
  https://www.inspire4.com